About the Authors

MAIA SZALAVITZ is the author of *Help at Any Cost: How the Troubled-Teen Industry Cons Parents and Hurts Kids,* which led to state investigations into the industry as well as federal legislation. She is a senior fellow at media watchdog STATS.org and has written for the *New York Times, Elle, Time* magazine online, and the *Washington Post.*

BRUCE D. PERRY, M.D., Ph.D., is the senior fellow of the Child Trauma Academy (www.ChildTrauma.org), a not-for-profit organization based in Houston that is dedicated to improving the lives of high-risk children, and he is an adjunct professor of psychiatry at the Northwestern University School of Medicine in Chicago. He is the author, with Maia Szalavitz, of *The Boy Who Was Raised as a Dog,* a bestselling book based on his work with maltreated children.

ALSO BY THE AUTHORS

The Boy Who Was Raised as a Dog: And Other Stories
from a Child Psychiatrist's Notebook:
What Traumatized Children Can Teach Us
About Loss, Love, and Healing

BORN FOR LOVE

Why Empathy Is Essential—and Endangered

Maia Szalavitz and Bruce D. Perry, M.D., Ph.D.

WILLIAM MORROW
An Imprint of HarperCollins*Publishers*

TO AARON, CELESTE, AND ELIANA SMITH:

May your empathy be a guide for others and
may the world grow more empathetic with you!

—Maia Szalavitz

FOR MY FATHER, DUNCAN RICHARD PERRY:

Miyotehew.

—Bruce D. Perry

FIRST HARPER PAPERBACK PUBLISHED 2011.

Library of Congress Cataloging-in-Publication Data has been applied for.

ISBN 978-0-06-165679-8

12 13 14 15 ID/RRD 10 9

We are all born for love. It is the principle of existence, and its only end.

—Benjamin Disraeli (1804–1881),
English statesman and prime minister

A human being is a part of a whole . . . [but] he experiences himself, his thoughts and feelings as something separated from the rest. . . . This delusion is a kind of prison for us, restricting us to our personal desires and to affection for a few persons nearest to us. Our task must be to free ourselves from this prison by widening our circle of compassion to embrace all living creatures and the whole of nature in its beauty.

—Albert Einstein

CONTENTS

AUTHORS' NOTE

There are many ways to collaborate on a book. Sometimes—as we did in our previous book, *The Boy Who Was Raised as a Dog: And Other Stories from a Child Psychiatrist's Notebook*—it makes sense to write entirely from one author's point of view. That book was based on Bruce's cases and traced his intellectual journey as a child psychiatrist through those stories.

For this book, however, we wanted to do something different. *Born for Love: Why Empathy Is Essential—and Endangered* is an exploration of ideas and stories shared and developed by both of us. That meant we had to write about situations in which one of us was present and the other was not. If we wrote it in Bruce's voice only, it would be hard to describe Maia's reporting—while if we wrote in Maia's voice, the same problem would occur in Bruce's cases.

Consequently, we decided to write as "we" and describe ourselves in the third person when just one of us was involved. That way, we could accurately detail scenes without eliding who was where and convey individual thoughts and ideas while retaining a consistent voice. We hope that this provides a satisfying narrative solution to the reader.

We'd also like to add a note here about the broad nature of this work. Throughout this book, we attempt to distill and describe some remarkable research from an immense variety of academic disciplines. Empathy is a truly universal topic. As a result, we had to be selective and even to oversimplify in some cases. We hope that our effort to summarize research did not distort the primary findings, key implications, or core principles of the important work of others. If so, we beg the forbearance of our colleagues. We hope that our readers—with whatever disciplinary background they bring to the subject—will read this book in the spirit in which it is written. We intend it as an exploration, a call for greater awareness, conversation, and broad debate about what we believe is our fundamental interdependence on one another and the crucial role of human relationships in the health of societies. We don't intend it as a final word, merely the beginning of what we hope will be a fruitful and truly transdisciplinary dialogue.

Finally, a note on names: those with asterisks (at first occurrence) are pseudonyms, and some identifying details have been changed to preserve confidentiality. In other cases, we used people's real first names only—this was their preference when they shared highly personal information. All other names are unchanged.

INTRODUCTION

S O WHY SHOULD I CARE? Why do people show concern for one another anyway? Are we really "born for love"? That's what this book is about: the empathy that allows us to make social connections, and the power of human relationships to both heal and harm.

There's been a recent explosion of scientific research on the subject, an incredible set of findings that show how empathy and the caring it enables are an essential part of human health. Indeed, one reason you should care—as the stories you'll read here illustrate—is that you'll live longer and be happier. But that's only the beginning. Empathy remains both intensely important and widely misunderstood. Its influence on the way we connect to each other can be seen everywhere, from the nursery to the Federal Reserve. And, as technology propels change at increasing speed, understanding the basis of these connections becomes ever more critical.

As we write, we're both bombarded by BlackBerry pings, cell-phone melodies, and Facebook notifications. We suspect that something important is changing in the way people relate to one another. We worry that the ability to empathize is in danger, that there might be, as a famous politician put it, an "empathy deficit" in America.

Bruce has long been astonished by the range of empathetic capacity that he sees in his work as a child psychiatrist. On one end is Ryan,* who we meet in Chapter 6, a young man from an excellent family, headed for a great college—who raped a developmentally disabled girl and boasted that he'd "done her a favor." On the other is Trinity, a woman who grew up to help hundreds of children in foster care—but whose early life was marked by abandonment and even murder in her family. Her story is in Chapter 7.

In between are most of us, including people like Eugenia (Chapter 3), who was adopted from a Russian orphanage into a loving family—but found that certain aspects of emotional connection don't come naturally to her. And Sam* and his son, Jonah,* who both have autism spectrum conditions that can cause a different set of problems in human interactions (Chapter 4). Throughout this book we'll meet folks whose lives illustrate conditions that can enhance or diminish empathy—and whose stories can help us understand how this one human capability can link us all across time and cultures, but may also be threatened by very specific situations and experiences. We'll also learn about cutting-edge science that is demonstrating ever more precisely how empathy matters to emotional and physical health.

Maia, meanwhile, has long wondered about empathy for more personal reasons. Oversensitive and bullied for being so as a child, she struggled with both an intense desire to connect and extreme personal distress that made her feel selfish when she couldn't. Why does it sometimes feel so hard to maintain friendships and family relationships? What do neuroscientists mean when they say we need one another to maintain the health of our stress systems, when they claim that our bodies are actually interdependent? Can you feel empathetic toward others—but fail to act that way because their pain overwhelms you?

We both are also concerned about the harsh tone of contemporary American culture: from calls for the legalization of torture to the actual practices uncovered at Abu Ghraib and Guantanamo Bay, and "torture porn" movies like the Saw series. There are also reality shows like *Intervention, Celebrity Rehab,* and *Brat Camp* and relentless attention to celebrity breakdowns like that of Britney Spears, which showcase other people's pain as entertainment.

Simultaneously, a worrying set of trends shows a measurable decline in social connection in America. For example, 80 percent of Americans say that the only people whom they feel close enough to confide in are family members. A full quarter say that they trust no one at all with their intimate secrets.[1] The proportion of people with no close friends or family members tripled between 1985 and 2004. Our trust in one another—an important factor in all types of relationships, personal and economic—has plummeted. Back in 1960, 58 percent of Americans endorsed the idea that "most people can be trusted"—but by 2008, this number was down to 32 percent (and it was already down to 33 percent by 1998, long before the economic crisis).[2]

Although online social networking is all the rage, real-world friendships and relationships apparently aren't doing so well. What is going on? Can we use our growing knowledge about the neurobiology of social life to understand it and help reweave the social fabric? This book is our shared journey to find some answers to these questions—and it all starts with understanding the way our brains connect us to one another.

This matters fundamentally because we live our lives in relationships. Shy or outgoing, rich or poor, famous or obscure—whoever we are, without connection, we are empty. Our interactions thrum with rhythm. From the moment of conception to the end of life, we each engage in a unique dance of connection. The themes and steps are shared by all humanity. They vary only in details and flourishes across culture, race, gender, and historical time. But they are inevitably shaped by those around us.

It is in our nature to nurture and be nurtured. However—as the bloody history of our species suggests—the development of these capacities isn't automatic. Empathy isn't extended to everyone. And certain specific experiences, certain particular actions on the part of those closest to us, are required for empathy to develop in children. Though Americans especially like to proclaim independence, our health, creativity, productivity, and humanity emerge from our interdependence, our history of relationships.

This interdependence is an inevitable product of our biology. For the naked, clawless, and not-exactly-fiercely-toothed human species to survive, we had to be able to form cooperative groups—small clans to hunt, gather, and collectively protect one another from starvation, predators, and, unfortunately, other human tribes. To reproduce and to keep our vulnerable infants alive, we needed one another. The resulting ability to read other people's intentions and to care about their plight—to empathize—helped us to become one of the most successful species on earth, the only one with the capacity to control its own environment. Humankind would not have endured and cannot continue without the capacity to form rewarding, nurturing, and enduring relationships. We survive because we can love. And we love because we can empathize—that is, stand in another's shoes and care about what it feels like to be there.

Mary Gordon, founder of Roots of Empathy, a children's program that we'll explore in the first chapter, likes to say that "Empathy can't be taught, but it can be caught." This book is about why we need an empathy epidemic. Empathy underlies virtually everything that makes society work—like trust, altruism, collaboration, love, charity. Failure to empathize is a key part of most social problems—crime, violence, war, racism, child abuse, and inequity, to name just a few. Difficulties with empathy or misperceptions of another's feelings also cause problems in communication, relationships, and business and are key parts of many psychiatric and neurological conditions like autism, depression, and antisocial personality disorder.

By understanding and increasing just this one capacity of the human brain, an enormous amount of social change can be fostered. Failure to understand and cultivate empathy, however, could lead to a society in which no one would want to live—a cold, violent, chaotic, and terrifying war of all against all. This destructive type of culture has appeared repeatedly in various times and places in human history and still reigns in some parts of the world. And it's a culture that we could be inadvertently developing throughout America if we do not address current trends in child rearing, education, economic inequality, and our core values.

So what is empathy exactly, and why does it matter so much? How does it develop—or fail to develop? Where does empathy come from in a world of "selfish" genes and competition to survive? How has it changed over the course of history—and what conditions allow it to flourish or cause it to wither? What can we do to get more of it? These are a few of the questions that have obsessed us: in Bruce's case, during his work as a child psychiatrist and neuroscientist, and in Maia's, during a career in science and investigative journalism.

WE ARE INDEED born for love. But at birth, we are not yet fully loving. Infants' brains are the most malleable—and vulnerable—that they will ever be outside the womb. The gifts of our biology are a potential, not a guarantee. As with so many other human potentials present at birth, empathy and love require specific experiences to develop. Just as Mozart could never have become a musical genius if his father hadn't provided lessons and instruments—and Michael Jordan would not have become the superb athlete he has been without access to hoops, balls, and courts—babies don't learn to care and connect without specific early experiences. Changes in the timing, nature, and pattern of these experiences will influence how relational capabilities emerge in an individual. These changes even help determine which of our genes will be activated and which will never reveal their potential—for good or for ill.

Humans have spent most of the past 150,000 years living in multigenerational, multifamily groups. These relatively small tribes were characterized by rich human interactions that aren't as present in developed Western societies. In these clans, the ratio of mature individuals to young children was roughly 4:1. That is, there were four caregiving individuals for every little one. Fathers, sisters, uncles, older cousins, aunties, and other kin surrounded children—and all of them could educate, discipline, nurture, and enrich. Two parents, many caregivers. That enriched social environment is what our brain expects.

In the modern era, however, the relational milieu has collapsed. In 1850 the average household size in the West was six people—today it's three or fewer.[3] A full quarter of Americans live completely alone. Hours and hours of television, educational ratios of 1:30 in school classrooms, mobile families, transient communities, nuclear families, broken families—all have contributed to reductions in the number and quality of relationships available to young children at the age when their relational needs are highest. Indeed, we now consider a ratio of one daycare worker to five children adequate! That is one-twentieth of the relational richness of a "natural" hunter-gatherer setting.

How does this change our connections to one another? What are the implications for our culture's capacity to care? Although our children are born for love, are we providing what they need to fully unleash that potential?

one | HEAVEN IS OTHER PEOPLE

I N A BRIGHT, AIRY LIBRARY at the West Hill Public School in Toronto, a class of seventeen sixth graders sits on the floor around a large green blanket, snapping their fingers. One delicate-looking girl wears a sky blue headscarf and matching top; another has dozens of long, thin braids, with strands of lavender threaded through her black hair. Every child seems engaged, focused.

Though it doesn't fit the stereotype of a violent, chaotic institution, this is nonetheless an inner-city public school. The eleven- and twelve-year-olds come from all over the world; this area of Toronto is a magnet for immigrants. Shannon Keating's class includes Native Cree children, African Canadians, a girl from Iraq, a girl from Afghanistan, a boy from Congo, and a child each from Poland and Guyana. Typically, many will move out of the transient housing they currently occupy before the school year ends. A significant proportion of West Hill students have spent time in homeless shelters.

But in the library with its sunny wall of windows, circled around the blanket, the children seem safe and secure. Even the boys are paying attention—fascinated by a tawny-haired, brown-eyed baby named Sophia, who will be six months old the day after this class. Sophia sits with her mom, Mary, calmly facing the sixth graders as they intently watch her.

The snapping intensifies as Sophia is placed on a cushion that allows her to lie on her belly, but still see the children around her. Each snap represents a synapse forming—the children have learned that Sophia's brain is growing faster now than it will ever grow again. Having been taught to observe Sophia closely, they are intrigued by how much they can discern about what she wants and how she feels just by paying close attention.

After about twenty minutes of introducing her to various toys and seeing her reactions, the class notices that Sophia's mood has changed. She starts rubbing her eyes and scratching herself. "Maybe she's tired," one child suggests. Mary picks her up. Although the theme of today's class was "crying," by attending so carefully to Sophia's signals, they've probably preempted it.

A Cree Native Canadian girl, dressed in shades of brown and black, is new in the class. This is her first day at West Hill. After the visit with Sophia, she is shy and tentative, seemingly confused. Two girls are talking nearby. One says to her friend, "It would be empathic to help her." So they do, explaining that they were watching a baby as part of a program called Roots of Empathy.

IF WE ARE all born for love, Roots of Empathy founder Mary Gordon was delivered into some of the most fertile ground imaginable. She grew up in Newfoundland, in a multigenerational household that included her three brothers and one sister, both of her grandmothers, and an uncle who was intellectually disabled. Her parents also often took in "strays." Unmarried women who'd gotten pregnant would live with them during their pregnancies, men leaving prison would visit nightly for a free meal. Gordon's father eventually served as the Canadian minister of labor, and her mother was an artist. The Catholic family was deeply committed to social justice. At the dinner table, the rule was that the conversation must focus on ideas—literature, policy, religion, philosophy—not gossip or mundane events. But the table rang out with laughter and spirited debate: this didn't produce sullen resentment.

Young Mary often tagged along as her mother visited the poor, bringing food, clothing, and coal to heat their houses, which were often dark, damp, and cold. If Mary made a face—and questioned why, for example, a woman might keep coal in her bathtub rather than use it to clean herself and her children—her mother would talk to her later. "My mom would never embarrass anyone, so she wouldn't embarrass me as a child, either. She saw the dignity in everybody. In the car she said, 'You judged that woman when you made that face.' She would say, 'She's made the best decisions she could make with the challenges she has. And you don't know her challenges.'"

On some weekends, she'd go with her father to visit the sick in the hospital, often reading aloud to them. Again, her father was careful not to embarrass those who might be illiterate: he'd always ask if the man they were visiting wanted books and if he'd like them to sit and read with him. And although all this charitable work might sound grim or self-sacrificing, Gordon's family infused it with pleasure and humor, so that's how the children perceived it, too. The visits were enjoyable social experiences—they liked accompanying Mom or Dad somewhere; it was fun, not a dull duty. Adds Gordon, "My relatives were all big storytellers. [And I learned that] literature opens the door to feelings and perspective taking. It's an invitation to be under somebody else's skin. I believe in the power of literature to elevate and connect us in humanity."

Now sixty-one, Gordon looks at least twenty years younger and brings an enthusiasm to her work that clearly links pleasure to caring.

With a kind, open face, a bright smile, and wavy reddish-blond hair, she's the type of person who can talk you into doing something difficult by making it look easy. She's warm and funny and makes compassion seem not just like the right thing to do, but the only thing to do.

Although she and her sister used to pray every night that they would not be called by God to be a nun, a nurse, or a teacher, all five of the family's children carry on their parents' mission in some way. One sister runs a children's choir that works toward world peace and has received the Canadian equivalent of a knighthood for her work; her brother is renowned author and journalist Gwynne Dyer.

Gordon began her work in the field of early childhood as a kindergarten teacher. "I couldn't believe it. Some of those little kids were so troubled already at that age, you wouldn't imagine that they could be so sad. People used to call them 'bad.' I thought what a name to put on a little kid. Every time they'd say 'bad,' I'd say sad. I decided that first week that this is not the most useful thing to be doing." She knew she had to reach families earlier.

Roots of Empathy was based on Gordon's observations of the way people responded to babies and what she'd learned at home. In 1981, she founded the first of many school-based parenting centers that have now become an international model for work with teen parents. She used to borrow a baby to carry when she was recruiting parents to participate. People with babies are more approachable; babies open conversational doors. "When we use little babies as teachers, it's not just the babies we're watching," Gordon says. "We're watching the baby in tandem with the parent. I believe that successful people develop empathy from receiving empathy or witnessing empathy."

There have now been nine independent evaluations of the Roots of Empathy program, including two randomized controlled trials. These found significant reductions in bullying and aggression and increases in "prosocial" behavior, including more sharing, helping, and inclusion of children who were formerly bullied or shunned. One study even found increased reading comprehension. The overall effects were long-lasting, too. They were still measurable three years later. Over 2,800 schools around the world now use the program, mostly in Canada, but also in New Zealand, the United Kingdom, and Seattle, Washington. More than 56,000 children were involved in Canada in the 2008–09 school year alone. Gordon has been asked to bring Roots of Empathy to the New York City public schools, but won't move forward until she can make sure that it is reproduced accurately.[1]

THE POWER OF the baby is clearly visible among the sixth graders. When Sophia smiles at them, even the coolest boys in the class break out in a big grin. This response is automatic: unless something went wrong during development, we all feel a stirring of joy in a baby's smile. And it's adaptive. If babies didn't produce some kind of pleasure for their older caregivers, they simply wouldn't make it. Infants are demanding creatures, often smelly, fussy, loud, and irritating. Something has to keep parents soothing, feeding, warming, and protecting them. They have to get some "reward" for meeting the needs of the infant.

Cuteness itself is one of Nature's tricks to get us to care for our young: the "awwwwww" response we have to small, vulnerable-looking things with big eyes is another way that babies seduce us into nurturing them. Think about how you feel when you see an adorable puppy or a fuzzy kitten: the impulse tends to be kind, protective. It's a unique emotion, one for which the English language has few words, and, until recently, there has been little scientific consideration of it. And it's no coincidence that smiling is one of the earliest developmental milestones, usually around four to six weeks, right when the reality of the difficulty of parenting begins to sink in.

Although cuteness may seem trite or silly, there's no doubt that it's appealing, particularly to females. Advertisers and manufacturers (and pets!) have long taken advantage of this evolved nurturing response to attract our attention because it is so powerful. It comes as no surprise to biologists that the Internet is overrun by charmingly helpless kittens and clumsy, huggable pups. Not to mention what must be the world's largest ever collection of baby photos and videos.

But what's going on in Baby Sophia's head when she smiles? Even moments after birth, babies can copy some facial expressions: for example, if you stick your tongue out, they'll follow suit. This ability is one of the earliest visible precursors to empathy. Another rudimentary form of empathy is the way crying is contagious among newborns: new babies will echo and amplify one another's cries, unable to distinguish another's distress from their own.

Although we call it "aping" when we copy one another, the capacity to reflect back the expressions of others is actually a highly sophisticated neural capability. Though some animals are "copycats" (curiously, probably not cats!), even nonhuman primates can't mimic actions with the precision and flexibility of humans. Nonetheless, this basic mirroring is the skill on which our ability to see the world through others' eyes—in other words, to empathize—rests.

The essence of empathy is the ability to stand in another's shoes, to feel what it's like there and to care about making it better if it hurts. The word itself was only coined in the early 1800s—it's a translation of the German *Einfühlung,* which means "feeling into." Sympathy—with which empathy is often confused—conveys something of the same idea and previously carried the meaning today given to empathy. The literal translation from the Greek root of sympathy is "feeling with"— and it is here that a subtle but important difference between modern thinking on sympathy and empathy arises.

When you empathize with someone, you try to see and feel the world from his or her perspective. Your primary feelings are more re-lated to the other person's situation than your own. But when you sympathize, while you understand what others are going through, you don't necessarily feel it yourself right now, though you may be moved to help nonetheless. Pity—or feeling sorry for someone—similarly captures this idea of recognizing another's pain without simultaneously experiencing a sense of it oneself. With empathy, however, you feel the other person's pain. You're feeling sorry "with" them, not just "for" them.

In coming chapters, we'll look at the hard science of this soft feeling, examining the components that are necessary for empathy and how they can go awry in various disorders and situations. We'll explore how empathy develops only under certain specific conditions—and what happens when a baby or a society doesn't experience these. We'll also look at how too much distress related to empathy can, in certain circumstances, be as problematic as too little—and at social influences on the expression of empathy. To get there, we first need to understand a bit about the brain, the stress response system, and how human contact can relieve or produce stress.

First, let's examine where empathy originates and how the ability to put yourself in the mind of another develops in babies like Sophia. Empathy is deeply rooted in our biology. The foundation for an ability to understand others starts with one of the most basic abilities shared by even the most primitive single-celled organisms. That is, the ability to distinguish self from other, your kind from mine. Even bacteria can sense the presence of others in their species—and more impressively, some can respond cooperatively to fellow organisms in certain situations.

This primordial self/other distinction arose from one of the most important challenges facing living creatures: to be a successful animal, you have to both survive and reproduce. You have to be able to know where you end and where others begin. In terms of sheer survival, of course, knowing what's likely to eat you, what's a part of you, and what looks like a part of you but is really a disease is crucial.

The immune system is the most obvious example of a mechanism to make these distinctions: its cells are designed to distinguish self from other and if "other" is detected, to kill. This can go horribly wrong in autoimmune diseases like multiple sclerosis, where the immune system mistakes your own nerve cells for invasive disease cells and attacks them. But most of the time, our immune cells protect us and kill invading microorganisms. They keep us separate, even when we are connected.

Outside the immune system, our senses need to detect friends and foes as well. And, in the most basic biological sense, the creatures most likely to be our friends are those with whom we share the most genes: our family, particularly our mother, father, and full siblings (you share approximately half of your genes with each such "first-degree" relative, unless you are an identical twin, in which case you share 100 percent with your twin). We're inclined to be kind to kin because our genes live in them, too—their survival represents our genetic immortality. One evolutionary biologist quipped that he would give his life for "two brothers or eight cousins"—either of which would statistically represent 100 percent of his genes.

This genetic preference produces the seeds of empathy. However, they can't sprout on barren ground. Empathy requires experience. Although we are genetically predisposed to care for others, the development of empathy requires a lifelong process of relational interaction. In our loving contacts, powerful genetic influences affect all of our biological systems—even permeating the most complex of human capabilities—language. And so, the root word for *kind* is *kin* and the word *kind* itself has the double meaning of being similar and acting in a caring and loving way.

Further, from the very start, the roots of empathy emerge from the soil of our stress response systems. We can see this as we watch Mary and Sophia: the attentive care that a mother gives her child shapes not just the brain systems involved in forming and maintaining relationships, but also the baby's capacity to "self-regulate," to control herself and her responses to feelings, thoughts, and experiences. However, if any experience is new or unfamiliar, it will initially be perceived as a "stressor," or a source of stress. When Sophia first visits the sixth-grade class, she alertly watches her mother for cues that tell her it is safe to interact with the other children.

The ability to respond to stress—and to control this stress response flexibly—is crucial to survival. Our brains have a widely distributed network of systems that receive input from both the external world of the senses and the internal world of the body. This stress response network is continually monitoring these sources of information for potential threats or needs for resources. For example, it has internal receptors that report whether the blood is low on glucose; if so, we start to feel hungry. It receives information from the eyes: if we spot a stranger with a gun, we feel fear, and adrenaline starts pumping, preparing us for "fight or flight." We rely on these stress response systems to keep ourselves as safe from as many kinds of danger as possible.

As adults, we can usually regulate these systems for ourselves. Although we will still always need at least some social contact to have a healthy life, we can cope with basic stresses solo. If we are hungry, we get something to eat; if we're in physical danger, we flee, hide, or fight back. We can learn to meditate, control our breathing, or go for a run. But an infant like Sophia cannot yet do any of this. She depends on Mary to feed her when hungry and keep her safe if threatened. Mary is Sophia's external stress regulator. It is this first key relationship—mother and child—that shapes the neural systems of the stress response to allow self-regulation. And it does so because the brain regions involved in relationships are the ones that modulate the stress response and allow empathy. These systems are interdependent. They develop together. This is one key to human connection.

Consequently, however, problems in the development of the stress response system can interfere with the development of social and emotional functioning—and vice versa. The brain's capacity to change with experience influences the way the infant perceives and responds to the world. The earliest, most fundamental experiences that shape the brain are these interactions with a baby's primary caregiver. They serve as a "template" that molds future responses to human contact.

This is one reason why empathy matters so much: from the start of life, we require others to help us cope with stress. Our brain requires social experience to develop properly: we influence each other's ability to manage stress in a very real, very measurable way. These connections are written into the architecture of our nervous system. If people are kind to us, our health tends to be good—but, as we will see, we can actually die from rejection and isolation. The fact that stress is regulated by social systems has tremendous implications for everything from medicine to politics to business to economics—and these make empathy essential for the survival of humankind. Since humans are a social species, this special mother-child dance is only the first of many—but it is the model that sets the rest of the relational machinery in motion.

And as the children in West Hill elementary observe, what goes on between mother and infant is not a one-way street. While Mary influences Sophia, Sophia also profoundly influences her mother. Babies naturally enjoy being imitated: if you want to get a baby's attention, one of the best ways to do it is to copy what she's already doing. The baby will respond by repeating the action, and if you add a variation, the baby will usually try to do that, too.

Watch yourself the next time you see a baby—if you aren't already doing this automatically, try it and watch how it excites and enthuses. In fact, you can engage with most young children almost instantly through responsive copycatting. A recent study confirms the social power of imitation: people prefer new acquaintances who subtly imitate their body language—and even capuchin monkeys prefer being with humans who mirror their actions rather than those who don't.[2] Subtle mirroring of certain body language is also a reliable clue to romantic interest.

In Mary and Sophia's case, the reciprocal nature of their relationship evokes pleasure for both mother and child. When Mary soothes Sophia, Sophia feels better—but so does Mary. When Mary makes Sophia smile, she feels joy, too. Sophia's smile has triggered something deep in Mary's brain. In fact, it activates a very powerful association—in essence, a memory—that Mary made years ago when she herself was loved as an infant by her nurturing caregivers.

How does this happen? How is it that we can get such pleasure from these brief human interactions? The answers to these questions reveal more about the fundamental power of relationships.

Any organism—from sea slug to human—needs some mechanism to ensure that its fundamental needs for things like food, water, and oxygen are met. So all organisms have a way to monitor the "levels" of these essential factors and to act to restore any imbalance. A serious imbalance causes distress, and correction brings a sensation of relief, often pleasure. The brain's networks for "reward" and "distress" have their origins in these primitive and essential regulatory systems. But if an animal doesn't feel hunger, for example, it won't eat and will starve to death. The brain senses and coordinates our responses to these needs. However, the brain isn't really just one organ. It's made up of multiple systems that evolved at different times, over which we have varying levels of conscious control.

Generally, regions that evolved earliest are located lower and deeper down in the brain. Sometimes called the "reptilian" brain because it is first seen in those creatures, the most primitive area is responsible for "automatic" functions like heart rate, which require moment-to-moment monitoring and immediate correction. These tasks are far too important to leave to the vagaries of consciousness and attention! Above the reptilian brain—and shaped during a later part of evolution—is the midbrain, which contains key areas involved in regulating sleep, appetite, pleasure, motivation, and attention. These regions also typically operate without conscious control. It is here that many of the stress response systems originate, sending direct connections down to the reptilian brain and up to higher areas, reaching throughout the brain. A region called the "limbic system" surrounds the midbrain—and this area is critically involved with relationships and emotion. The most "advanced"—or at least uniquely human—parts of the brain are the highest and outermost brain regions, those of the cortex, which allow language, abstract thought, and planning.

Importantly, the widely distributed architecture of the stress response network allows it to "take over" any parts of the brain needed to respond to a threat, including the "thinking" cortex. Also important to note is that these brain regions work in concert, so it is impossible to actually separate "rational thought" from emotion. Even the most sophisticated decisions and analyses require positive or negative emotion; otherwise, it is impossible to determine which choice or idea is "better" and which isn't. Valuing anything—even an idea—as "good" or "bad" requires feeling.

Most of our brains' nonstop monitoring and responses to our fundamental needs take place outside of our awareness. For example, if you are sitting in Shannon Keating's classroom with Maia and your brain senses a slight shift in the ratio of oxygen to carbon dioxide, it will signal your lungs to take in more air. You might yawn or take a deep breath, unaware of why you did so.

Alternatively, imagine that you are hiking in the mountains of Colorado with Bruce. The combination of strenuous exercise and thin air produces a much more severe oxygen/carbon dioxide imbalance. The normal response doesn't work: soon, you find yourself panting, but your brain still feels like it isn't getting enough oxygen. Now you become conscious of the problem. You start to feel discomfort, distress, and probably a little anxiety. Your stress response system is telling you something is wrong.

These signals soon go up to the most advanced part of the brain, shifting your attention away from the scenery or your plans for next Tuesday. The stress response system can literally shut these cognitive regions down. It's as if it is saying, "Stop thinking about how beautiful the damn mountains are and solve this oxygen problem." So you pause. You find a place to sit and rest. You soon start to feel better and, interestingly, now find yourself able to truly appreciate the grandeur of the view.

It turns out that even before you became consciously aware that you weren't getting enough oxygen, your stress response system was affecting your ability to calmly contemplate the world around you. It is much harder to use the higher systems of the brain—to think of the future, to be creative, or to be nurturing—when you are in distress. This fact about the stress response is of great importance in terms of our ability to empathize—and is another factor in how people can influence one another in both helpful and hurtful ways.

The brain itself has many primitive but powerful ways to motivate us. It gives us pleasure when we do things that reduce distress, engage in activities that increase our chances of reproducing, or take actions that improve safety for ourselves or our children. Under the opposite circumstances, if we feel unsafe, have some unmet hunger, or are unable to protect or nurture our offspring, it causes pain or distress. These connections between pleasure, stress, relationships, and our ability to think creatively will be explored throughout this book. But for now, just keep in mind that a sense of safety is rewarding, while threat is distressing, even painful.

At first, when Sophia is just a newborn, her pleasure comes simply from relief of the distress of being hungry or cold. But soon, through the hundreds of times when Mary attends to Sophia's needs—feeding her when hungry, warming her when cold, and comforting her when frightened—Sophia's brain begins to connect all Mary's attributes with that comfort, satisfaction, and pleasure. These cues become associated with a decrease in the stress response: a positive interaction with Mary makes Sophia feel safer. These crucial associations between positive human interactions, reward systems, and the stress response networks are the neurobiological glue for all future healthy relationships. They are at the core of why empathy matters.

Mary also gets pleasure and relief when she is able to ease her child's hunger by feeding her. For both mother and child, this pleasure is dual: it is not just the baby's relief that comes from the cessation of the pain of hunger (or, for the mom, the end of the distress of hearing her baby cry) but also, for Sophia, the pleasure of the taste of the milk itself, of being held close and smiled at. For Mary, feeling her baby mold into her body, touching Sophia's soft skin, smelling the indescribably lovely scent of her baby's head, hearing her coos, and seeing her adorable face also brings happiness. So both types of pleasure—the kind that comes from satiating a desire and the kind that comes from physical experiences like good tastes or warm touch—are combined.

Soon, Mary doesn't even need to start feeding Sophia to elicit joy. She lights up and stops crying simply seeing Mom come into her bedroom. It's hard to explain just how rewarding that feels to most mothers—many describe it as being comparable to, or even better than, the happiness of reciprocated romantic love. Suffice it to say, it feels pretty amazing to be able to make someone ecstatically happy just by showing up. This, of course, helps both Mary and Sophia through the many difficult moments of development—and the positive cycle of social reward spirals onward.

All this mirroring is also reflected in the brain: quite literally in a set of recently discovered cells called "mirror neurons." As noted earlier, babies are born with two important skills that prepare them for empathy—the ability to begin to imitate facial gestures and the automatic response in which the cries of other infants cause them to cry as well. Both of these appear to involve mirror neurons. These cells—discovered by Giacomo Rizzolatti and colleagues in the early 1990s—have revolutionized our understanding of how we understand one another.[3]

Mirror neurons fire when you do something—but more important, they also fire in a less intense fashion when you see someone else do that same thing. So, for example, if you see someone smile, your mirror neurons respond. These cells essentially copy the pattern of activity you would experience if you were smiling, but without fully completing the muscle movement. And, indeed, invisibly but measurably, the muscles in your face involved in smiling do respond slightly, even if you are just looking at a photo of a happy person.[4] These neurons show you "what it's like" to experience what others do. Similarly, if you see someone cry, your mirror neurons respond as well. This allows us to feel one another's joy—and pain. Quite literally: brain scans show that when people see others in pain, some of the same regions of the brain light up as when they experience pain themselves.[5]

Although he is best known as the economist of the "invisible hand" of the market, Adam Smith was one of the first to note the importance of this emotional experience in morality. He presciently described what we now suspect to be the effects of mirror neurons: "When we see a stroke aimed, and just ready to fall upon the leg or arm of another person, we naturally shrink and draw back our leg or our own arm; and when it does fall, we feel it in some measure, and are hurt by it as well as the sufferer."[6] That feeling, he said, is the basis of compassionate action. Such empathy is also the foundation of trust, which is necessary for the successful functioning of everything from relationships to families to governments and, yes, to economies.

Sages, religious leaders, and philosophers have, of course, long known that consideration of others is a cornerstone of morality. In all great religions, for example, there's an equivalent of the Golden Rule, a summary of moral teachings that suggests that considering how you would want to be treated in the same situation is a good guide to doing the right thing. From the biblical "Love thy neighbor as thyself" to the Taoist "Regard your neighbor's gain as your gain, and your neighbor's loss as your own loss," to the Islamic "None of you [truly] believes until he wishes for his brother what he wishes for himself," and the Talmudic "What is hateful to you, do not to your fellow man. This is the law: all the rest is commentary." Each of the world's great religions expresses this basic principle. Socrates, Epictetus, and other Greek philosophers also articulated it; Kant's "categorical imperative" expands it to consider the effects of one's action on others not just in the particular instance, but as if your action were to become the basis of a universal law. All of these "Golden Rules" show how greatly morality depends on empathy and our ability to see the world from other points of view. And this starts with mirror neurons.

It's as though we were born with a program that automatically runs a simulation of the experience of others. Rather than having to try to consciously consider, "What would it be like to feel what he does?" we do it without even thinking, our brain mapping the other's experience onto our own limbs and body. This kind of empathy isn't at all about conscious deliberation. We experience it like any other sensation, and choices come in only later, when we decide how to act on the information.

Babies like Sophia, however, can't yet determine exactly where their bodies end and those of other people start. When infants cry in response to another newborn's cries, they feel distress. But they don't really recognize that it is another baby who is upset. Nor do they fully understand that that baby isn't part of them and has a distinct mind of her own. They just experience "something bad"—and start to cry. At about six months, as they gain greater control over their bodies and their responses to the world, most no longer automatically cry if another child does.

Now, if Sophia heard another child wailing, she'd probably whimper and look disturbed, but she wouldn't automatically cry herself. She has begun to see herself as a distinct being. Since she doesn't experience internal signals of distress from her body as she would if the pain were her own, she experiences less discomfort, which is easier for her to manage. At this age, self-control begins to develop. Just as babies first cry instinctively when they feel pain or hunger, as their brains grow, conscious control over crying develops. Crying no longer "happens to" them—it can, of course, if pain is sharp or sudden enough, as it can, indeed, in adults—but over time infants learn that they can choose whether to cry or not.

As they grow, young children start to become aware of other people as distinct entities as well. If Mommy looks sad, for example, a toddler might bring his blanky to her—because toddlers know that when they are sad, they like to cuddle the blanky. Between the ages of one and three, children also begin to show other spontaneous helpful behavior: if they see an adult struggling to do something that they know how to do, they'll often demonstrate it, without being asked. Toddlers will also mirror their parents' distress or calm: you've probably seen little ones this age look to Mom or Dad to decide whether or not to cry after a fall. Mirroring guides the growth of empathy, choice, and self-awareness. It gives us a natural way to understand and care about others.

And as they learn to speak, children become even more sophisticated about the distinction between self and others. At age one, if Sophia sees her friend Emily crying, she will seek her own mother for help, even if Emily's mother is nearby as well. But by two and a half—interestingly, around the same time children begin to recognize themselves in the mirror—if Emily is upset, Sophia will probably find *Emily's* mother for her. Sophia by then would recognize that one's own mom is the one you want for comfort, not someone else's. Around the time they turn three, children also begin to learn that their parents aren't omniscient—that other people know different things from what they know, based on what they've seen and heard. Chapter 4 examines how problems with understanding these kinds of questions about what other people think and feel are a big part of autism spectrum conditions.

In this way, the normal development of empathy begins, first as a nameless feeling and a set of inborn responses, with no distinction between self and other. Next, awareness of others grows, as does awareness of your similarities and differences. Early empathy is just emotional contagion—the spread of a feeling from one person to another, without any sense of separateness. One baby cries—soon the whole nursery is bawling.

The ability to sense the emotions of others soon develops into the knowledge that other people have separate bodies and experiences, too. The parent/child dance sets the pace, creating templates that shape and color later relationships. It is here that empathy is first taught or—as Mary Gordon puts it—caught.

MAIA RETURNED FOR a second Roots of Empathy visit between Shannon Keating's sixth graders and Mary and Sophia four months later, toward the end of the school year in 2009. Sophia had grown—as had her connection to the class and to Mary. Now much more independent, Sophia stood and held on to her mom, looking around with her huge brown eyes as her wide smile lit up the whole room. By observing her, the class had learned that crying doesn't mean a baby is "bad" or "doesn't like you," and that people have different temperaments. As one child put it, he'd found out that "Everybody is different in their own way. Like maybe somebody is athletic, and some people may not be athletic, but everybody is in their own way unique and special."

The class had also watched how Sophia learned new skills over time, seeing that not being able to do something now doesn't mean you won't ever be able to do it. This kind of information and their emotional connection to Mary and Sophia seemed to help them be more compassionate toward both others and themselves.

When Maia talked with the class, they had just completed an exercise in which they had to choose several words to describe themselves. Shiva, who is from Afghanistan, said, "My words are 'wonder' and 'empathy.' I picked 'wonder' because sometimes I myself wonder who I am and 'empathy' because I care about others and how they feel and I try to make them feel happy. I think Roots of Empathy is not only good for children but I think people who are older should know about this too so then they learn how because some people never realize how important it is to have empathy."

Said Cleopatra, the African Canadian girl with the long, skinny braids, "To me, empathy means when you can understand what someone feels, and relate, and comfort them as well—just to be able to know how they feel."

Yasmine, also African Canadian, wearing her hair in short pig-
tails, raised her hand and said, "I learned a lot of new things. I feel
proud to know that I'm helping the baby's neurons connect and so her
brain can grow. I learned that thing from Mary Gordon, 'love grows
brains,' so I think that's kind of catchy to remember." She added, "It
shows you how even the smallest person who seems—like Sophia
doesn't speak with words—she seems insignificant but she makes the
most impact."

Maia asked Yasmine—who continually amazes her peers and
teachers with her intelligence, vocabulary, and insight—what empathy meant to her. She replied, "Empathy is the ability to understand
another person's feelings and metaphorically step into their shoes
and understand that sometimes if someone is acting differently, maybe
it's just because they have something on their mind. So maybe you
should listen to them to see what they're going through and feeling."

SECONDS AFTER JEREMY* WAS BORN, there was a sudden hush in the tumult of the delivery room. Despite the delirium of labor, his mom, Angela,* immediately noticed the pause and asked sharply what was wrong. But soon, she saw for herself. Her baby boy had been born with a facial defect, a large, dark hairy spot that nearly covered his right cheek. An ugly thing with an ugly name, she would later learn: a hairy nevus.

Angela resolved then and there that no matter what anyone else said or did, she was going to show her son that he was beautiful—and make up for the stares and taunts that she knew would come his way. She would love him, she would make it OK. The world might have a hard time looking at him—but Jeremy would know from the start that his mom adored him exactly as he was. That was the right thing, the only thing, for a good person to do. She empathized with her child, deeply. The idea that any harm might ever come to him was physically painful for her to even consider. She felt an overwhelming wave of love for this tiny child, who had so recently been part of her. She'd never felt anything like it before.

Her husband, Terry,* was amazed by her acceptance and ease with the baby. He soon came to share her intense love for their little son. At first, he hadn't been sure how to deal with it, but pretty quickly, Jeremy's face simply seemed normal. Terry didn't flinch or think anything about it when he played with Jeremy or lifted him out of his crib. They made arrangements for surgery to remove or, at minimum, reduce the defect, but that would take time and multiple operations. They knew the birthmark was something they'd have to cope with for most of Jeremy's childhood, at least.

Jeremy's family came to Bruce's clinic when he was four years old. Angela wasn't convinced that their son needed to see a child psychiatrist, but Terry insisted. The last straw for him was the preschool fiasco. Angela had carefully selected a preschool, making sure the staff understood Jeremy's special needs. On the first day, however, when Angela tried to leave, he'd thrown such a frightening and unremitting temper tantrum that both the school and the parents agreed that Jeremy wasn't ready to start.

But would he ever be? As Bruce talked with each parent in succession, it became clear that Jeremy had no tolerance for any kind of discomfort. If something didn't go his way, he screamed—and usually Angela fixed it for him immediately. He couldn't take a second of frustration without blowing up. He had to have his way. Terry said that he had become worried that Angela was spoiling him—but she had told him that Jeremy needed all the extra protection and support she could give because of his face.

That tantrum in Target after they'd refused to buy him a toy? That was really the result of strangers staring at him. When he insisted on sleeping in their bed? Jeremy needed that comfort to deal with the challenges he faced from other people's reactions to him every day. His preschool meltdown? The other children hadn't accepted him and that one aide had looked at him funny. In trying to buffer the world's negative reaction to Jeremy's face and by relating everything that happened to his birthmark, Angela had instead prevented her son from

learning to cope with stress. To help him, we'd have to help the whole
family understand the relationship between caring, stress, pleasure,
and empathy.

AS WE SAW with Sophia and Mary, during normal infant development,
mother and child become attached to each other in a reciprocal connection that links pleasure with soothing each other and happiness
with making each other happy. Here's how it works: when a mom runs
to soothe her crying baby, she is actually tapping into a set of memories from her own early childhood. If the mother was cared for in a
loving way, her brain made associations between her own mother's
touch, gaze, smile, and other characteristics and pleasure. So now,
many years later when she calms her own child, these actions stimulate a set of key neurotransmitter networks in her brain.

Triggering this "memory" causes the release of several important
neurotransmitters, which are chemical messengers in the brain. These
particular transmitters are linked with pleasure and were first understood in relation to drug addiction. The first chemical involved is dopamine, which is connected to a sense of desire and "wanting." This pulls
the mother toward her child—the obsessive drive addicts have to seek
drugs is an exaggeration of this bonding mechanism, which originally
evolved to link parent to infant. The second set of relevant chemicals is
actually a group of small protein neurotransmitters, called "endor-
phins" or "enkephalins." These are the "endogenous opioids," aka the
brain's own private heroin. And they produce the pleasure, content-
ment, and relaxation that mother and baby enjoy together. All of this
makes the mother's nurturing actions calming and pleasant.

So now the mom's stress response system, having been activated
by mirroring that of her distressed infant, quiets down. The same thing
happens to the baby. The release of endogenous opioids and dopamine

is an aspect of the stress response, a part of the cycle that helps re-store the system to balance. Though the nature of this reward system can make people vulnerable to pathologies like addictions, most of the time craving affection is natural and healthy, as is the interdependence it creates. There has to be a biological way to ensure that we will con-nect with others: these chemicals and the pleasure they produce are the glue that bonds us.

But opioids and dopamine aren't the only important chemicals in-volved. When a mom nurtures a baby—or when anyone acts to form a kind and trusting connection—another protein is also released. It's called "oxytocin" (not to be confused with the painkiller Oxycontin!). Inside the baby's brain, dopamine, opioids, and oxytocin become ac-tive as he settles down in his mom's arms. Their bond is forming.

Oxytocin is necessary for mammals to make the connection be-tween a particular individual and pleasure. Without it, many animals can't even tell each other apart. Its work starts early, when mother first connects to child. Its chemistry seems to be essential to empathy and to all the relationships that depend on it. By itself, oxytocin appar-ently isn't psychoactive and doesn't produce any kind of high or joy. However, it is released in both males and females at orgasm. And there, too, it seems to link pleasure and particularity. The joy of sex? That's down to opioids and dopamine. But passion for your own true love is parsed out in oxytocin (although for males, another hormone is needed as well, called "vasopressin").

Once that connection is made, however, being away from your be-loved becomes unpleasant. It's like a milder experience of heroin with-drawal. Separation produces craving, but in this case, for a person, not a drug. Our parents are the first people with whom we experience this intense bonding. And it happens even in baby rats. In laboratory experi-ments, researchers have found that giving opioids is one of the few ways to relieve the separation cries of rat pups taken from their mothers. Conversely, giving drugs that block the effects of opioids makes these cries more intense. Interestingly, oxytocin works, too, even if opioids are blocked. That means that oxytocin has soothing powers beyond those of opioids alone. (The mechanism is still unknown, but it may involve dopamine.) Of course, as everyone knows, being reunited with Mom is best of all. The pleasure and relief in this reunion is plain to see.

But how do we know when someone else is experiencing pleasure? What gets humans to mirror a state of happiness? Typically, we look for signs of joy in someone's face. And the reasons for this take us back to the origins of humankind. During the thousands of generations of the early history of our species, we lived in small bands composed of roughly forty to sixty members. Individual survival depended on close cooperation and communication. Back then, and even today, much of human communication did not take place through language. Messages were sent and received through subtle cues in posture, expressions, and tone of voice. The face is the most important of all of these nonverbal social communication "instruments."

Facial expressions reflect our moods and feelings. In turn, they can elicit a specific emotional and social response: a smile, frown, glare, or snarl, a message of "come hither" or "get lost." Although we evolved basic facial expressions whose meanings are universal, their nuances can be culturally specific. A genuine smile is a signal of goodwill worldwide, for example—but a head shake may be read differently in various cultures. And even more specific nuances in our inclinations toward particular interpretations can come from our own early experiences.

If children have a nurturing set of caregivers from the start, they will build up a personalized catalog of familiar faces, those of their family and friends. These are the stored templates that come to mean "familiar/safe." By seeing and responding to these known faces, children learn the nonverbal language of their culture, even before they pick up spoken words. Later in life, an unfamiliar face—even in a non-threatening context—will elicit a low-level alarm, a small activation of the stress response system.

This is because all new faces are judged by the brain as potentially threatening. One reason for this is that, in general, the brain is conservative. You are more likely to survive if you assume an innocuous garter snake is a poisonous rattler than you are if you make the opposite mistake. Consequently, we see most new things as threatening until proven innocent. Another, more powerful reason that new faces elicit at least some sense of alarm is that we evolved in a world where for thousands of generations, the major threats to any individual were other humans, typically strangers. Unlike other animals, the most lethal predators we faced were our own kind—not members of different species.

Early in human history, seeing a new person, a new face in an interaction, meant that there was another clan nearby—competing for the same water, fruits, game, caves. This new person was as likely to attack you, drive you away, steal your camping site, take the young, and rape the women of your band as he was to decide to negotiate and cooperate. Across generations, wariness of new individuals, groups, and ideas was built into the circuits of the human brain's alarm response because those who had this wariness were more likely to survive to reproduce. It was just safer to assume danger—to expect the worst—than to count on the kindness of strangers.

So through thousands of generations of evolutionary selection, the brain developed the capacity to read nonverbal cues, many of which are communicated via changes in facial expression. The brain has special face and expression recognition capabilities, which should be the envy of the Department of Homeland Security. There's a whole region of the cortex devoted to recognizing faces called the "fusiform gyrus." Through a process of "matching" expressions and faces with prior experiences of them, the brain makes decisions about threat and safety. There are a host of interesting implications to this—including potential insights into how this neurobiological tendency for tribalism can contribute to group conflict, racism, nationalism, and a set of other unhealthy, nonproductive human characteristics. We will examine some of these in later chapters.

For now let's focus on how the face is a critical source of information about relationships, a key conveyor of data needed for empathy. Simply smiling can produce a small neurochemical "reward." When we see someone else smile, a little connection is made. A "smile" can also be seen in the eyes—we know by looking into someone's eyes whether they are smiling genuinely or just giving us a polite grin. Only the real smile produces pleasure. But we can only get these signals face-to-face.

WITH A BIRTH defect like Jeremy's, however, the pleasure of sharing smiles and eye contact can be disrupted. Because his face is so different from the "catalog" of safe and familiar faces that most people have stored, they will tend to respond with a look of horror, disgust, or fear. When people meet babies with facial defects, they usually do try to respond appropriately, but it's hard to cover the initial response, and the result is rarely a fully reciprocal smile. Instead, there's a forced grimace or stilted grin. The child's brain perceives this confusing and deceptively signaled emotion. As Lucy Grealy, a writer whose face was disfigured by treatment for childhood cancer, revealed, "I learned the language of paranoia: every whisper I heard was a comment about the way I looked, every laugh a joke at my expense. . . . I was my face, I was ugliness."[1] A baby, of course, can't conceive of such issues—but seeing other people's negative reactions, not surprisingly, can spur upset, rather than connection.

Alternatively, as in Jeremy's case, his mother's admirable determination not to let the defect have any impact can yield a different set of problems. His early experience wasn't primarily one of others' negative reactions—but it wasn't one of normal mirroring, either. And that interfered with his ability to cope with stress. Understanding how can show us the process that normally allows people to modulate one another's stress responses.

And it all starts with the face. Ordinarily, during nursing or bottle-feeding, mothers (and fathers, too, but we'll use the example of mothers here) and babies spend hours looking into each other's eyes. It is through this contact that they synchronize with each other, the mirror neurons producing imitation that both the mom and the baby then elaborate as they react to each other playfully. Not surprisingly, oxytocin has been found to increase the time people spend looking into each other's eyes. Normally, this interaction is rhythmic and flexible. Mom looks at baby, looks away, and moments later, reconnects with her gaze. Baby looks at Mom, hears a noise and turns, comes back. (In blind children, these patterns are established through voice interactions and responses.) But little breaks in contact are essential—they are basically small experiences of minor stress and distancing, quickly ended by reconnection. The stress response systems are shaped by these little breaks, the child ultimately learning to manage small, repeated doses of stress activation without overreacting.

It's important to note that stress itself is not bad; in fact, all learning involves at least a small dose of stress because it requires exposure to something new and unfamiliar. Through these experiences of tiny stresses, the brain develops the capacity to manage moderate and larger ones. By learning that food always comes eventually, wet diapers will get changed, Mom and Dad do come back, a child also discovers that mild distress can be tolerated and will ultimately be relieved.

In fact, frequent activation of the stress response is necessary for learning and healthy development. The key to a healthy stress response capability—to healthy self-regulation and ultimately resilience—is the pattern of this stress. When stress response networks are activated in small, moderate "doses," they become stronger over time. In contrast, large, irregular, or extreme doses can interfere with their development. They need the right kind of "exercise" to develop strength and resilience. Only with thousands of repetitions of little doses of stress do these neural networks become "strong." Fortunately, healthy development involves innumerable minor stressors starting with these brief breaks in soothing face-to-face engagement with a parent.

With baby Jeremy, however, Angela was determined not to look away. Other people would do that—not Mommy! She wouldn't turn away from him ever. She had quickly become accustomed to how he looked—her initial shock was replaced by acceptance and familiarity. She smiled genuinely and looked constantly, but she didn't allow the normal rhythm of detachment and reconnection to become established. Even though Jeremy would break eye contact, seemingly trying to establish a normal rhythm, Angela didn't respond to his cues. She was literally always "in his face." Even as he began to crawl and walk and play on his own, she hovered.

If she was on the phone and he called for her—that was the end of the phone conversation. If she had friends visiting and he got frustrated with a toy, she'd run over and fix the problem, telling him it was hard and she would help. Even with Terry, it became clear that Jeremy was the first priority—which began to affect their marriage. Although it came from a wonderful motivation, all this constant focus didn't give Jeremy's stress system a chance to mature, to move away from being modulated completely by his mother. Though it is impossible to "spoil" babies by being too responsive when they cry or by giving too much affection when it is sought, never letting children experience frustration and ignoring their attempts to disconnect can interfere with their capacity to cope with stress. Jeremy didn't get the small, repeated moderate stress doses followed by soothing connections that would allow him to progress toward comfortably handling larger stresses and separations. And Angela tried to do away with larger stresses, too, as much as possible. Consequently, Jeremy's stress system didn't mature past the low frustration tolerance and lack of regulation of infancy. He couldn't regulate himself—and so he couldn't begin to recognize the needs of others. Without self-regulation, he wouldn't be able to fully develop the capacity to empathize.

To understand why this should be so, we need to know a bit more about the brain and how it develops.

THOUGH WE ALL know how helpless newborns seem, many people are unaware of just how immature their brains are. Other animals—like horses, for example—can walk just after birth. Many animal babies are rapidly equipped to do most of what adults of their species can. But because the human birth canal is so narrow, much of the brain development that other animals undergo in the womb occurs outside of it in humans. In essence, all human babies are "premature." Being adapted to walk upright provided many advantages. Nonetheless, it also limited the size of our newborn babies such that they come into the world with extremely undeveloped brains. Consequently, human infants cannot walk or even crawl or feed themselves without help.

They grow fast, though: during the first three years of life, infants' brains undergo intense change. In fact, if the body developed at the same rate that the brain does, we'd be trying to change the diapers of five- or six-foot-tall toddlers! Ninety percent of brain growth takes place in these early years. The parts of the brain that are first organized and functional in newborns are the regions concerned with regulating life's necessities like breathing, swallowing, and eating.

Development then proceeds from these central brain areas located toward the bottom of the brain upward and outward, roughly following the order in which the various regions evolved. This means that the lower, more central areas are the most primitive, while the higher, outer regions mediate our most advanced functions like language. As the higher regions develop, they gain some control over the lower areas. Nonetheless, even in adults, threat or distress shifts control away from the rational, abstract thinking areas to the "more decisive, rapidly acting" central, lower regions. Under perceived threat we get dumber but faster, which can help us survive in a fire or when fleeing from a bad guy, but can also get us in trouble at work or in other social situations.

This sequential development of the brain, from lower to higher, is matched by parallel advances in physical and linguistic skills. Most of us are familiar with these landmarks—crawl, walk, run, throw, or, for language, cry, whine, babble, word, sentence (though "whine" tends to stick around a bit longer than most parents would like!). The sequential mastery of our motor, emotional, and cognitive functions continues throughout childhood as the relevant brain areas mature. The last part of the brain to mature—in fact, not fully developed until the early twenties—is the cortex, which allows self-control, complex thought, and foresight. (And some cortical regions don't fully develop their fastest connections until the thirties!) If you've ever been or parented a teenager, you know how late these regions come online.

All of these various areas of the brain need to be in interconnected to allow for smooth orchestration and integration of the brain's many functions. One of the most important architectural features of the brain is a set of neurotransmitter systems that originates in the central, lower areas of the brain. These systems include norepinephrine, dopamine, and serotonin neurons that send projections up to virtually every part of the brain and down to the autonomic nervous system that regulates the heart, lung, gut, pancreas, skin, and rest of the body. This wide distribution and central location (right where all the sensory input from the outside world and the body come in) allow these systems to regulate and influence the rest of the brain and body. It is no surprise, then, that these are some of the most important components of the stress response system.

Another key component of the stress response is the hypothalamic-pituitary-adrenal (HPA) axis. The brain networks that control the HPA axis are also deep in the brain and are regulated by the neurotransmitters mentioned above. Through the HPA axis, stress-related hormones such as cortisol can be released to help the body and the brain adapt to impending threat. In later chapters, we'll explore how experiences that interfere with both of these important stress response components can lead to pervasive problems for both the brain and the rest of the body.

And as we saw with Mary and Sophia, it is the attentive, attuned, and nurturing care of a baby's primary caregiver that begins to shape and regulate these developing stress response systems. The foundations of healthy self-regulation are developed during the first few years of parent/child interaction, the years when Jeremy's mom became so intensely focused on providing nurture without stress. Angela's actions were well intentioned—but unfortunately they were "out of step" with the kind of stimulation that Jeremy's brain was expecting. For the stress system, no stress can be as bad as too much stress: the key is moderation and rhythm. Patterned, repetitive activity is, in fact, necessary to all kinds of learning—whether building the neural systems needed to manage stress or trying to build stronger muscles.

Just as you wouldn't build muscle by resting all week and then trying to lift a hundred pounds just one time every Friday morning, you can't build a healthy stress response system by complete protection from stress or occasional exposure to an overwhelming dose of it. A healthy stress response system is exercised regularly with moderate, manageable experiences—and so when it does later encounter large stressors, it is much better equipped to cope. Normal parent/child interactions provide these small, manageable doses of stress in a pattern that creates resilience. And, for all of this, the face is an essential source of information.

ALTHOUGH THE DEVELOPMENT of the full capacity of empathy relies on parents, the need for others to maintain a healthy stress response system isn't limited to childhood. Throughout life, we need social contact to regulate our response to distress. Of course, exercise, meditation, and many other stress relief techniques can be done alone—and periods of solitude can help reduce the stress that relationships themselves can cause. But in the absence of any close human connections, nonsocial stress relief tactics can rarely sustain health. To take an extreme example, being in solitary confinement is one of the most distressing experiences someone can have. Normal people can rapidly become depressed and even psychotic if completely isolated for periods as short as a few days. Long-term isolation can raise blood pressure, lower immunity, and worsen virtually all mental and physical illnesses. As University of California psychologist Craig Haney—one of the world's leading experts on the destructive effects of isolation—put it, prisoners in solitary often suffer "an impaired sense of identity; hypersensitivity to stimuli; cognitive dysfunction (confusion, memory loss, ruminations); irritability, anger, aggression, and/or rage; other-directed violence, such as stabbings, attacks on staff, property destruction, and collective violence; lethargy, helplessness, and hopelessness; chronic depression; self-mutilation and/or suicidal ideation,

impulses, and behavior; anxiety and panic attacks; emotional breakdowns and/or loss of control; hallucinations, psychosis, and/or paranoia; overall deterioration of mental and physical health."[2] Nearly 90
percent develop "irrational anger," and another researcher found that
a third suffer "acute psychosis with hallucinations."[3] Though people
also benefit from choosing reflective solitude—even, at times, for
long periods—we need relationships to be healthy.

And so even as adults, most people find simple acts of social contact like holding hands soothing. When we're sick, we often still
yearn for Mom and Dad, and though any familiar person can have a
calming impact, a mother's touch is often still the most regulating
presence. Holding a loved one's hand—or even just seeing his or her
face nearby—can lower blood pressure and levels of stress hormones.
As adults, we can do this for one another, usually without our own
stress getting out of hand. Getting to that point as an adult requires
the complex early duet between parent and child, where little stresses
are experienced, then relieved. But unfortunately, that wasn't what
was happening in Jeremy's case.

BRUCE WAS GOING to have to teach Jeremy and his family this basic
information about the development of the stress response systems if
Jeremy was ever going to be able to tolerate preschool and later educational and social experiences. Starting would be difficult because the
amount of distress Jeremy could tolerate was extremely limited. Even
small things could easily prompt a long-lasting tantrum, from which
he might take an hour to recover. The only way forward would be in
small steps—to give him the tiny doses of stress and then relief his
system needed to "practice" and, ultimately, to mature.

Because he'd had such erratic experiences of stress, Bruce gave
Jeremy a little chemical assistance, with a medication called "clonidine,"

originally used for controlling blood pressure. This medication acts
on the norepinephrine system, a key driver of stress regulation in the
brain. By damping this system down, clonidine can help prevent the
spikes of overactivation that can lead to disruptive behavior and difficulty returning to calm. This would help the boy begin to regulate his
behavior and stress reactivity.

Bruce also talked with both Angela and Terry, helping them to
understand how the stress response system works and how they could
use this information to help Jeremy. He made sure not to appear as
though he was blaming them for the problem: sometimes, the best-
intended acts don't have the desired effects. The parents' motives were
good—their compassion had an unintended result because it wasn't
balanced by knowledge of the positive effects of small challenges, of
facing and mastering small stresses. In these situations, it's always
important to recognize the parents' strengths and minimize shame
and guilt. When parents feel blamed, they are less able to empathize
with their children's problems and may respond defensively, instead of
changing their behavior. Focusing on the solution and on how difficult
it is to know what to do in such cases helps families move forward. So
that's what Bruce did.

The next step would be to provide very small doses of stress and
relief repeatedly, to help Jeremy develop the capacity to tolerate a little
distress. As with muscles, gradually increasing patterned, repetitive
activities makes them stronger—while massively overloading them at
random times can injure and weaken them. Bruce started by having
Angela provide Jeremy a slow introduction to preschool.

First, he would only have to spend five minutes there on his own.
Angela would explain to her son that she would only be away for a very
short time, just long enough for her to go to the bathroom and come
back. When he was able to tolerate that without crying, she left him
for ten minutes, to go to the store to buy a soda. Jeremy was told what

to expect at each increment, and each step was repeated consistently until he could manage it. Consistency and responsiveness are also essential to regulate stress—for all of us, in fact. Soon Angela was able to go away for a whole hour to have lunch while Jeremy played with the other children without melting down. He began to interact and share with them more. And, after a few more increments, he could tolerate a normal preschool day.

Another area of conflict for the family was bedtime: Jeremy still couldn't fall asleep in his own bed and had to spend every night with his parents. It was driving Terry crazy. This, too, was handled by gradually increasing Jeremy's comfort level, step by tiny step. Before long, he was able to sleep in his own room. As Bruce worked with the family, Jeremy was also undergoing a series of operations to remove the hairy nevus. This, too, couldn't be done all at once because the skin grafts to replace it had to be given time to grow in. Sometimes major stressful experiences like these surgeries would cause Jeremy to regress. But over time, he became a sturdy and dependable little boy who went on to thrive in elementary school—both socially and academically—and beyond.

ONE OF THE key factors in Jeremy's situation was that his birthmark was on his face. We don't usually think that much about it, but the way we respond to the world—and the way people respond to us—is largely played out on the stage of the face. People with facial defects or with paralysis that affects regions of their faces can be deeply affected by this. When the world views your face with horror or disgust, it's difficult not to feel rejected and isolated. Alternatively, if your face is paralyzed and doesn't respond to the smiles or other signals of others, they may feel rejected or slighted by you, causing distance and discomfort in the relationship. Emotions themselves can become muted when they are not part of reciprocal communication. People with facial paralysis from conditions like Parkinson's disease can come to seem dull and boring—even if they are actually engaged and excited—because their faces cannot participate in this signaling. Our faces display our emotions—but they also shape them.

A terrible irony here is that empathy itself may help create the high levels of disgust many people experience when seeing people with facial injuries or defects. Empathy and disgust, in fact, are mediated in part by the same brain region—and having a high level of one may be linked with having high levels of the other. One reason people seem to respond instinctively with horror to malformed faces is because they imagine that having the defect is painful or that it must be unbearable to have others respond to you by flinching. The distress that this produces in the viewer may be so intensely upsetting that they respond not by being kind or reaching out to the victim, but by avoidance. The emotions aroused by feeling deeply empathetic may actually sometimes prevent a genuinely kind and empathic response!

Research confirms these observations that overempathy can sometimes backfire. One study looked at children aged five to thirteen who viewed distressing videos of children being separated from their parents or being wrongly punished, and scenes of disabled children having difficulty climbing stairs. The more distressed the victim was, the more upset the children watching became—and the more focused they were on the victim and how to help. But this was only true up to a point: when the children became more distressed than the victims appeared to be, they turned the focus inward, to try to help themselves feel better, rather than considering ways to help those who were hurting.[4]

Another study, which had kindergartners play the roles of children who were sick or in pain, found that those who measured highest on empathy beforehand were least likely to help others after they had played these roles. By imagining themselves being hurt or ill, they became extremely distressed. Then, they were too upset by their own vicarious experiences to reach out and soothe others who were suffering.

Adults can have the same problem. One study found that the most empathetic nurses were most likely to avoid dying patients early in their training, before they had learned to deal with the distress caused by empathizing too much.[5] Overempathy can look from the outside like selfishness—and even produce selfish behavior. In Jeremy's story, we can see that at times, you can have too much of a good thing.

E UGENIA SITS NEAR HER BED, in her sunny, ordered bedroom. Unlike in many teen abodes, the bed is made, no crumpled laundry is on the floor, and no stray shoes have escaped the closet. The eighteen-year-old's collection of Russian nesting dolls and miniature animals is dust-free, placed carefully on uncluttered shelves or atop a dresser near a mirror. Large windows look out over her leafy suburban Massachusetts neighborhood. In her lap on an orange towel are her two large guinea pigs, Wally and Stanley, which she strokes gently as they purr.

She's alternately enthusiastic and shy, bubbly and easily embarrassed. Her strawberry blond hair is long, perfectly straight and shiny; her brown eyes provide a striking contrast to her freckles and light hair. After she speaks with Maia, she will drive to the local animal shelter, where she volunteers almost every day, usually helping with dogs or rodents. A shih tzu puppy, now six months old and adopted from the shelter, runs in and out of her bedroom, each time demanding attention and bringing in a different squeaky toy.

"I don't miss people," Eugenia says, looking up and smiling nervously because she knows that others consider this odd. "I guess I don't get attached to people, pretty much. I feel a lot closer to animals."

——

EUGENIA WAS ADOPTED from a Russian orphanage when she was two and a half. After seeing Bruce discussing the effects of early childhood neglect on the *Oprah Winfrey Show,* she got in touch with him for help with a school project. The *Oprah* episode had explored the most extreme cases of neglect: those of "feral" children who were either raised by animals or not parented at all, a subject we discussed in our previous book, *The Boy Who Was Raised as a Dog.* Eugenia had a personal reason for doing her presentation on feral children. Not only had she suffered neglect herself at the orphanage, but she had recently learned that she had a sister who lived with and was friends with one of the feral children featured on the show. That child, Oksana Malaya, had been raised in a kennel with dogs for much of her early life by her abusive alcoholic parents. When she was discovered by authorities, she walked on all fours, barked, and panted like a dog. Although she did learn to speak, she remains institutionalized, with profound social and psychological problems.

Eugenia wanted to understand how early neglect affected people— and when we told her we were writing this book, she wanted to share her story to help others learn more about the development of empathy. Though her case is nowhere near as extreme as Oksana's, the type of neglect that she suffered is unfortunately far more common. About nineteen thousand children are adopted by Americans from foreign countries each year, and nearly 90 percent of these children have spent time in an institutional environment that can be considered neglectful because it fails to meet the developmental needs of young children.[1] Nearly half a million cases of neglect by parents are also confirmed in the United States each year.[2]

Eugenia herself is kind and gentle; she's not cold or unconcerned about social matters. But her early experience has left some marks that she has had to work to overcome—and she would probably struggle a great deal more with empathy if her adoptive parents hadn't provided her a stable, loving home. Understanding just what goes wrong when children don't get individualized care in their first several years can help us recognize what's needed for those years to go right.

AS WE'VE SEEN, the ability to handle stress normally develops in concert with a baby's relationships with her primary caregivers, usually parents. But the connections being built here are very specific. Baby Sophia, for example, doesn't initially learn a general principle like "people will help me" or "being kind is good" by being nurtured by Mary (though later, of course, these kinds of generalizations can be made). What she does learn is that a few individuals with unique scents, voices, faces, and styles of touch are there for her. She associates these specific cues with safety and comfort. For Sophia, safety means Mary's smell, touch, and look, not someone else's. "Just anyone" with hands, food, and clean diapers won't do. Learning these associations takes numerous repetitions. With each interaction, the bond with that particular person is strengthened.

By about eight months, babies have usually built strong "memory templates" of their bonds with the people who appear in their daily lives. Now being approached by strangers starts to cause fear and crying—they have built in enough of a catalog of who is familiar and safe for them to recognize when they don't know someone. Eight- and nine-month-olds are notorious for not wanting to be held by anyone other than Mom—or whoever their primary caregiver is. This intensely tight bond rarely extends to more than one person—and will only extend even partially to others if those folks have been with the baby pretty much every day.

Although babies this age can recognize many familiar people, typically, they cannot easily be soothed by anyone but their primary caregiver. Secondary caregivers like Dad or Grandma or a big sister might do for a while, but Mama is who they really want. Fathers often feel rejected by the strong preference that babies show for their mothers at this age and feel as though they've failed when they can't soothe them. But this is just a normal stage in the development of attachment, which gradually grows to accommodate more people as the child's brain becomes ready for this complexity.

In Eugenia's early life, however, this bonding process was disrupted. For reasons that are unclear, when she was about two or three months old, her mother left her at an orphanage, near where they lived in Russia. Her adoptive mother, Sue, says that Eugenia's birth mom, Galina, is a "lovely woman who has had a very hard life, and it shows." Galina appears to have been abused as a child and possibly had an abusive relationship with Eugenia's biological father. That father apparently didn't help in his daughter's upbringing, and he didn't stop her mother from placing the child in an orphanage.

In some parts of Eastern Europe, orphanages are unfortunately sometimes used as babysitting services. Dropping a child off at one, at least in some social groups, is nowhere near as stigmatized as it would be in America today. It's not necessarily seen as permanent. Consequently, though Eugenia's adoptive parents were notified about her case when she was eighteen months old, they were forced to wait another year to bring her home to be sure her birth mother wouldn't reclaim her. That was one more early life year she would spend in what we now know to be a very damaging environment.

To baby Eugenia, the situation was baffling. Having been fed and nurtured by her mother for her first few months, she suddenly found herself in a world where she was just one of dozens of needy babies. Mommy was gone—as was everything and everyone else she knew in the world. Although people don't have conscious memories of infancy, babies' early experiences nonetheless deeply imprint themselves into the brain. In fact, there is growing evidence that during pregnancy and in the first few months of life, babies' brains are more vulnerable to environmental influences than at any other time in life.

As they say in computer programming, however, this isn't a bug, it's a feature. By probing the world around them for signals about the availability of resources, babies' developing brains begin to express the biological settings best suited to that environment. For example, if a baby senses—due, perhaps, to being underfed—that food will be hard to come by, her brain may express the genetic settings that preserve every calorie possible. If she is actually born into a world of plenty, however, that child is at risk of becoming overweight. In Eugenia's case, however, the orphanage was short on emotional, rather than material, resources. So that's the world for which her brain prepared.

To those unfamiliar with an infant's developmental needs, the orphanage looked wonderful. The rooms were bright and open, there were plenty of toys. Babies and toddlers were moved from one activity to another: they weren't left alone in their cribs. Staff monitored their play. Trees and grass surrounded the building; there were safe places to play in. And everything was clean and sparkling. "You could eat off the floor," Sue says. Eugenia was expertly changed and fed by a rotating staff. On the surface, it didn't look especially troubling.

But, in fact, it was a loveless void. In the orphanage, Eugenia wasn't special to anyone anymore. As Sue put it, the children were simply "herded" from one activity to the next, not played with as individuals. Eugenia didn't receive intensive, particularized adult attention and affection. To the staff, she was a job. Even though many may have truly cared and enjoyed working with children, she was still just one of about a hundred babies. When she returned years later, although some of the women who'd worked there in her infancy were still around, no one really remembered her. One or two said her name sounded familiar. Such forgetting, of course, would be unimaginable for a parent, short of dementia. That's the depth of the bond that was broken.

AND, AS AN INFANT, Eugenia spent her nights alone in a crib, her days mainly in a large, padded playpen with about ten other babies around the same age. Though she looked cute and was affectionate, she didn't grow at a normal rate. Among others raised in the same lacking environment, of course, this wasn't at all obvious. As Sue and her husband, Ken, discovered when they got her home, Eugenia had also been infected with rare parasites.

It has long been known that children living at orphanages are more susceptible than others to most illnesses. At first, infectious disease was blamed. After all, germs and infectious agents of all types spread most efficiently in places where large numbers of people live in close quarters. Consequently, once germ theory was understood, great efforts were made to keep orphans' rooms sterile and to isolate the children from one another. Physical contact with staff that might spread germs was also limited. As researchers later discovered, this actually made the problem worse. Even now, orphanages continue to be dangerous and sometimes deadly to babies and toddlers.

What's going on and why does empathy matter here? The first person to study the issue systematically was René Spitz, an Austrian physician and psychoanalyst. Although he had no way of understanding the mechanism, he suspected that children in institutions were suffering from the loss of important relationships—and that connection between parent and child was critical to health. He didn't think that infants could survive, particularly psychologically, without that special attachment. He tried to understand the situation from the child's perspective. Soon, he and another key theorist, John Bowlby, would begin to help the world recognize the importance of parent/child bonding—and the attachment we now know to be so critical for so many aspects of health.

In 1945, Spitz compared babies raised in the then-typical sterile American orphanage with those raised in a cold, institutional prison nursery. In the orphanage, there was one nurse for every eight babies; there were sheets between the cribs in an attempt to minimize contagious disease. All day, the orphans lay alone, receiving little attention or affection. In contrast, the prisoners' babies spent most of their days with their incarcerated mothers. If the institutional setting itself and its inevitable germs was causing the children's bad health, the two situations should have had identical outcomes for babies. Alternatively, if what infants really needed most was individualized parental attention, the children in prison should do better. Spitz couldn't randomize babies from one setting to the other, but he figured that there were unlikely to be important distinctions between the groups other than maternal care.

And indeed he found massive differences. They were glaringly obvious without even studying anything more subtle than the death rate. Thirty-seven percent of the children in the orphanage were dead before they reached age two—whereas none of the infants raised in prison with their mothers died.[3] Lack of individualized care wasn't just unhealthy, he'd discovered; it was frequently fatal, killing more than one in three infants subjected to it. Though such a staggering infant mortality rate is common in situations without modern sanitation or medical care, it was not at all normal in industrialized countries, even six decades ago.

The prison children did better in every way that Spitz could measure. They contracted fewer infections, were healthier in general, and grew at the appropriate rate. They seemed intellectually and emotionally normal. In contrast, the orphanage children were often sick, didn't gain weight on schedule, and showed other signs of profound emotional disturbance, cognitive delays, and disabilities. The empathic connections the prisoners' babies made with their mothers literally saved their lives: lack of love was what was killing the orphanage children, not unsterile group living.

Nonetheless, over half a century later, Eugenia and thousands of other babies were—and sadly, many still are—being kept in settings where they don't receive individualized care. When Eugenia was adopted, she was in the tenth percentile for weight and height: she was tiny because, as we now know, individualized and frequent physical affection is needed to just spur growth hormone production. This is true for all mammals. When "runts" in litters of puppies and kittens die, these deaths are typically caused by the same problem. Runts tend to be too weak to stimulate their mothers to give them enough grooming and attention; they can't get access to or suck hard enough on a nipple to get noticed and nurtured. Without being nuzzled and licked, growth hormone isn't released. Even if a runt does then manage to fight his way to a nipple and eat, the food isn't metabolized properly and he "fails to thrive." Without the physical experience of "love," the body senses little chance for survival and shuts down. In humans, "failure to thrive"—basically, a cessation in normal growth and development—is a frequent consequence of orphanage life and other severe forms of neglect.

Unfortunately, many childcare officials around the world still do not take this evidence into account and continue to believe that orphanages are safe for infants. Consequently, a more recent and properly randomized study was published in 2007 to prove to the Romanian government that its orphanages were damaging children. Almost as starkly as Spitz's research had done, it demonstrated how important ordinary empathetic parenting is to the development of virtually all important brain functions.

Following a ban on abortion in 1966, Romania experienced an epidemic of infant abandonment. The abortion ban at first doubled the birth rate, but draconian policies like stationing police in hospitals and requiring monthly gynecological exams of women of child-bearing age did not make unwanted children into wanted ones. Until the mid-2000s, the country didn't have a foster care system for infants, so these babies languished in orphanages. In fact, Sue and her husband, Ken, would adopt a younger brother for Eugenia from Romania in 1994. Researchers utilized the terrible Romanian orphanage system itself to try to put an end, once and for all, to the critique of early childhood deprivation research that claimed that children will develop just fine without individualized attention and affection.

There had long been criticism—from behaviorists and others—regarding the early childhood deprivation research. For one, it wasn't randomized—so there could have been preexisting differences between the groups of children. Some critics argued that the reason orphanage-raised children do so poorly is not because individual parenting is necessary for human development, but rather because healthy babies get adopted and the rest have birth or genetic defects that only look like they were caused by institutionalization.

Since there were not yet enough foster homes for all of the infants, the babies in the Romanian study were randomized either to be put in foster care or to stay in the orphanage. Those kept in the orphanage were still eligible for adoption—so the study didn't mean that any children were deprived of individual parenting who would otherwise have received it. Children with obvious medical problems and disabilities—like Down syndrome or fetal alcohol syndrome—were excluded from the research. The foster care and institutionalized children were also compared with children born at the same hospital who stayed with their biological parents. Those who have blamed birth defects for the poor outcomes of children raised in institutions cannot make that argument about this research.

In terms of IQ alone, the results were striking: at forty-two months, the lucky children in foster care had IQs nearly nine points higher than those who stayed behind at the orphanage. The average IQ of an orphanage child was 77, whereas it was 86 for the foster care children. Those who had never been institutionalized had IQs of 103, on average.[4] The difference between the foster care and orphanage children might not seem massive, but when you consider that the standard cutoff point for mental retardation is 70, those points can really matter. And even the children in foster care hadn't been placed there at birth: on average, these infants had already spent their first twenty-one months of life in an orphanage. That's nearly two years—at a time when the brain is at its most vulnerable. Even for something as seemingly unrelated to emotional development as IQ, empathic care is necessary for optimum development.

The study also clearly showed that the younger a child was when he was removed from the orphanage, the better he did both socially and intellectually. If researchers had compared babies adopted at birth with those adopted later, we suspect the difference would have been many, many times larger. The institutionalized children were physically smaller and had smaller head circumferences than the others—a very blunt measurement of brain development.[5] Bruce's work on similarly neglected children had earlier shown the same thing. He found both dramatically smaller head size and reduced development in brain areas that were supposed to be expanding at that age. These differences are so dramatic that someone completely untrained in brain imaging—even a child—can easily see them.

Animal research adds to these conclusions: monkeys raised in isolation without mothers are socially inept, to the point where they can't even mate successfully. They are also anxious and hostile. If such females are inseminated, they often kill their babies. Interestingly, however, if allowed to spend time with their infants, the neglected monkeys get better at mothering with each successive litter, showing that although early deprivation is damaging, the worst outcomes can be mitigated.[6] Fortunately, this is often true for humans as well.

EUGENIA'S NEW PARENTS, Sue and Ken, brought her home when she was a toddler. In the early 1990s, although research already strongly suggested the dangers of early neglect, little had been done to educate the public about it. "When I look back, we were pretty ignorant," Sue says. "We were just committed to the thought that kids need a home, we have a home, and we were very idealistic that love can conquer all. Now that I've raised two kids who were institutionalized for a relatively long period of time, I will tell you that there are definitely issues and we have worked really hard to overcome them."

Sue is the CEO of a national company; her husband runs a software firm. This gave them access to the best schools and other resources for their children. Although their careers are important to them, family has always come first. Throughout their children's lives, at least one parent has worked from home. The couple decided to adopt a few years after they got married when they discovered they were unable to conceive. "I wasn't really comfortable playing with mother nature," Sue says. "I felt as though if it was meant to be, it was meant to be, and I wasn't about to start loading my body with drugs."

Sue comes from a large Italian family and had always wanted to share that joy by becoming a mother. She wasn't going to let infertility stop her. In fact, her major regret about how she's raised her own children is that she wasn't able to live near extended family, the way she grew up. During Sue's childhood on Long Island, her aunt and maternal grandmother lived next door, cousins were everywhere, her paternal grandparents lived nearby, and the whole family gathered every Sunday for a big Italian feast. Sue says she based her values on those of her mother. She was not only a rock in her own family, but a support for Sue's teen friends, some of whom remain in touch with her mother, even as adults. "She understands that life is relationships and you give unconditional love, she's loyal, she's understanding, and is very empathic," Sue says. Sue's mother passed down these qualities in the way she nurtured her daughter—giving Sue the maternal tools she needed to nurture Eugenia.

Ken and Sue learned about Eugenia from a representative of an international adoption agency. After several other international and domestic adoption attempts fell through, they finally brought their own baby girl home from Russia. She was tiny—and very active. She spoke Russian—a language in which she's still fluent, thanks to tutors and classes to maintain it. Although she was very affectionate, she was also constantly in motion.

"She was a climber; she was real hyper," Sue says. "The first year you had to really watch to make sure she didn't kill herself." She points at a high mantelpiece in their well-appointed living room, saying that if she'd turned her head, she'd find that Eugenia had climbed to a seemingly impossible and dangerous height. The little girl picked up English rapidly, however, and soon adjusted to life in America. Once the doctors figured out how to treat the rare parasitic infection she'd contracted at the orphanage, she began growing quickly. About a year and a half later, Ken and Sue adopted a younger child, Bobby, a two-year-old boy from Romania.

Ken flew to Romania to pick Bobby up. One of Eugenia's earliest memories, in fact, is waiting at the airport to meet her new brother, when she was not tall enough to see over the bar that separated the waiting area from the disembarking passengers. Her parents gave her a Snow White doll in hopes of convincing her that having a baby brother would be fun—but they soon found that she didn't need much convincing. Rather than being jealous of the potential interloper, Eugenia became his translator and advocate. Whenever he appeared to be asking for something, Eugenia spoke up and explained to the adults what he wanted.

Oddly enough, however, she probably could not have verbally understood what he was saying. Sue originally thought that Bobby was speaking Romanian—a quite different language from the Russian that Eugenia did know. As it turns out, he wasn't speaking any known language: either it was gibberish or a language the orphans had developed themselves in his orphanage. Though it sounds unlikely, this is actually reasonably common. Although language must be learned, the human drive to communicate is powerful. There are quite a few cases where children in institutions or whose parents speak different languages have created new tongues. Twins, too, often develop their own languages—though these are not usually as complex.

Eugenia was a very observant little girl: she watched her brother carefully and learned to read him. This attentiveness probably helped in her own development of empathy. By being there to tell her parents what Bobby wanted, she learned to consider the needs of someone outside herself. She loved her little brother and played frequently with him. The age gap between them probably felt smaller to her than it would have otherwise because previously institutionalized or otherwise neglected children tend to bond better with younger boys and girls. Even though they can catch up surprisingly quickly in loving homes, they tend to seem younger than their chronological age.

Eugenia drank in the praise her parents gave her and the comfort she could see she brought to her brother when she was able to express his needs. "Maybe I was just guessing as to what he wanted," she says.

"It was kind of strange, but I always understood. I could say he wants a drink or whatever."

Conditions in Bobby's orphanage had apparently been much worse than in Eugenia's. Unlike his sister, Bobby wasn't toilet trained when he was adopted, and he had difficulty with motor skills. His leg muscles were severely underdeveloped. Sue thinks he may have spent most of his infancy lying in a crib, not being given the chance to crawl or walk much—which could account for this problem. Bruce has seen the same difficulties in other children whose neglect included restricted ability to move freely. Muscles and the brain regions that control them both require exercise in order to work properly and develop; without this, they atrophy. Bobby has required years of occupational therapy and uses a keyboard rather than writing by hand because of difficulties he still has with fine motor control. However, Sue says his emotional adjustment was easier than Eugenia's when they first brought him home. Eugenia's presence and her energetic advocacy of her brother may have helped him adjust.

Sue and Ken never hid the fact that their children were adopted. Both were given age-appropriate information as they asked. Eugenia, however, clearly began to yearn for her birth family as soon as she was able to begin to express the thought. "I don't look like any of you," she'd say to her parents. In elementary school, she became fascinated by twins and spoke constantly about wanting to be a twin herself. She became a huge fan of the Olsen sisters, Mary-Kate and Ashley, idolizing the twin stars and obsessively seeking information about them. Her parents bought her tickets to one of their concerts for her eighth birthday. Now Sue wishes that she'd realized earlier that there were psychological needs her daughter was expressing through this passion. She wanted a blood relative.

———

SO WHERE DID Eugenia's yearning—this "hole in her heart," as Sue calls it—come from? How, ordinarily, do mother and child actually bond? To understand this—and discover more about the roots of empathy—we can look to the relatives of Eugenia's purring guinea pigs: the prairie voles. It might seem a big leap from orphanages to prairie vole burrows, but these garden pests actually have a lot to teach us about the biological basis of empathy, parenting, and even romantic love.

Voles are small, mouselike animals, sometimes called field mice. If you've heard of them, you might have heard them called "love rats." Outside of the scientific and pop-sci literature on affection, they are best known for destroying lawns and gardens by burrowing into roots and eating grass. If they aren't eating your lawn, they are quite cute, big-eyed fuzzy brown balls of fur with shorter tails than house mice.

And if you want to understand love, empathy, and social behavior, voles can be your best friends. Because as we've learned from the voles, the chemistry of monogamy and that of mothering are quite similar. In fact, oddly enough, the brain systems that allow mothering may have evolved and expanded into the neural basis for monogamy. It was only after studying monogamy in voles that the brain basis of maternal bonds first began to be understood.

The man who first recognized that there might be something special about prairie voles was field biologist Lowell Getz, now professor emeritus at the University of Illinois. In the early 1970s, he noticed that the same male and female prairie voles kept turning up together in his traps. He never dreamed just how important this observation would turn out to be. As he put it addressing a well-attended 2009 conference devoted entirely to vole research, "Those of us who do basic research are almost constantly asked, 'What good is it?' and so you say, 'Well, you never know.' This little information you're finding may sometime have some application."[7] Literally thousands of scientific papers on topics including sex, parenting, monogamy, aggression, and empathy would ultimately result from Getz's esoteric study of the love life of an unassuming field mouse. This research offers key insights into why social contact is so important to human health, how early childhood wires social connections into our brains to cope with stress, and how, in humans, empathy becomes possible. It is in large part due to vole research that we've begun to understand the physiology of how Eugenia's early life may have affected her brain.

Getz recognized immediately that his observation of vole pairs was worth further investigation. If vole couples were spending significant time together, it could mean that they "pair bonded," which was not typical rodent behavior. Most rodents—including lab rats, mice, and guinea pigs like Eugenia's—mate promiscuously or have harems. They don't settle down as couples to rear young. Whereas 90 percent of birds pair bond to raise nestlings together, only 3–5 percent of mammals do.[8] If prairie voles proved to be one such species, their behavior definitely warranted study.

And so Getz and his team of assistant "Mouseketeers" put tiny radio collars on prairie voles. They discovered that vole couples did indeed live together and raise young together, the males doing everything short of nursing to nurture their little ones, as well as vigorously defending their nests. Getz and his colleagues would ultimately study voles for twenty-five years, putting down 2,320,300 traps and handling 236,700 of the little creatures.[9] (No word on how often they got bitten!)

When he was sure he was seeing monogamy, Getz turned to a colleague, C. Sue Carter, to study the behavior of these unusual rodents in the lab. Unlike many biologists of the time, Carter realized that there had to be a physiology that drove monogamy, that something chemically must be different in prairie voles compared with other rodents. In other words, this behavior had to be based in a particular part of the brain: it wasn't likely to rely only on general learning.

Carter was already studying reproductive hormones and behavior. So she knew that in 1971, researcher Peter Klopfer had discovered that oxytocin was involved in mothering.[10] Oxytocin had previously been identified by British pharmacologist Henry Dale as a cause of uterine contractions during labor. In fact, even today, a synthetic version, called Pitocin, is used to induce labor or speed its progression in hospitals around the world. Dale named the hormone for this ability: oxytocin comes from the Greek for "fast birth." But Klopfer went further. His research, he believed, suggested that oxytocin was what he called the "hormone of mother love."[11]

That made sense to Carter. Nature was often parsimonious: why shouldn't a hormone involved in birth and nursing also help bond mother and baby? But Carter thought it might go even further. She wondered if oxytocin could be involved in other social behavior—perhaps in the monogamous relationships of these voles, in addition to their parenting activities. At the time, however, most scientists didn't believe that oxytocin was active in the brain.

Over the years, Carter and her colleagues began to unravel the chemistry of connection. And it would help explain not just monogamy and motherhood—but trust and friendship and stress relief, too, as we shall see. As it turns out, other species of voles are not monogamous: meadow and montane voles mate promiscuously; their males are cads, not dads. Unlike prairie voles, montane vole males don't stick around and help with the young. (If you ever want to keep your voles straight, just remember that Prairie voles are NOT Promiscuous and Montane and Meadow voles are NOT Monogamous.) The fact that such similar species showed such different behavior was a boon to science: if differences were found in their brains, they would most likely be the source of the behavior the researchers were interested in.

So what was going on inside their mousy little heads? As Carter—now known as the "mother of oxytocin research"—suspected, oxytocin was a key part of it. Early experiments showed that injecting it into rats' brains made them behave more socially. These loved-up rats were also less fearful around other rats. As we'll see throughout this book, love is physiologically designed to conquer fear—but there are circumstances that can prevent this.

Further research revealed that female prairie voles had a high concentration of oxytocin receptors in brain regions involved in the experience of pleasure. This pattern of receptors was quite different in promiscuous female montane voles: few were located in pleasure regions.[12] It was amazing; these females looked like they were literally "wired" to desire and get pleasure from one specific mate. The oxytocin connected the experience of being with their particular partner— that guy and him only—to craving and pleasure. Female montane voles, however, lacked these connections and could want and get pleasure from any male.

The way this wiring is laid down is fascinating as well. Carter and her colleagues discovered that if a virgin prairie vole female hangs out with a male and mates with him for long enough, her brain changes radically. Prairie voles typically go at it in repeated bouts for twenty-four hours when they first meet—one mating is not enough to trigger the change that leads to bonding. But giving oxytocin—even if the voles just chill and don't mate at all—will do it.

As noted earlier, oxytocin itself doesn't seem to be responsible for the actual sensation of pleasure during sex. It just connects the pleasure regions—which release desire-linked dopamine and satiation-linked endogenous opioids—with a particular partner's scent. If you block oxytocin or block dopamine in voles, no bonding occurs. In fact, blocking oxytocin in rodents produces "social amnesia"—they can no longer even tell one another apart, let alone figure out which guy or gal is their sweetheart. But allow oxytocin to flow, and prairie voles soon find themselves hooked on a one and only.

Further investigation showed that while oxytocin did aid bonding in both sexes, in males, another chemical was needed as well. Vasopressin—originally known as a hormone involved with controlling urination—seems to be critical for male parental and monogamous behavior. As with oxytocin in females, in male prairie voles, vasopressin receptors are concentrated in dopamine- and opioid-rich brain regions. This is not true in their promiscuous cousins. And similarly, repeated mating by a male with one partner wires these regions to respond with pleasure to her scent and with pain to her absence. Vasopressin also makes prairie vole males aggressive against intruders: once mating triggers a vasopressin/pleasure connection, they become fierce defenders of their nest, as well as nurturing dads to their babies.

Interestingly, however, monogamy in voles is rather like monogamy in humans: far from perfect. About 25 percent of baby voles are not genetically related to the father who shares their nest.[13] Clearly, the female here is stepping out. As with humans, however, "don't date married people" remains good advice. In the lab, only about one-third of females will pick up and move in with a new "lover" and abandon their "husband."[14] Two-thirds stay, even though some have dallied. Even in rodents, even among our best animal model of monogamy, relationships aren't easy or simple!

Monogamy is, however, ineluctably connected to motherhood. Oxytocin researcher Larry Young, professor of psychiatry at Emory University, suggests that monogamy may have evolved when maternal nurturing and bonding initially directed at babies somehow expanded to include bonding with sexual partners. It's strange to think that mothers could have "fallen in love" with their babies long before they were able to fall in love with their sex partners. However, although monogamy seems to have evolved at several different times, it never did so in species that don't nurture their young together. Paternal bonding could have followed the same pattern. Though Freud postulated that our psychology is shaped by an unconscious desire to mate with our opposite-sex parent, this theory rather turns that one on its head! If Young is right, in evolutionary history, maternal love for babies came first.

SO HOW DOES oxytocin work its magic on mothers? Most women who have been given it to induce labor in the hospital find the stuff unpleasant: the pain from intense contractions isn't typically endearing. However, that reflects the actions of oxytocin on the uterus—not the brain. Nursing mothers often describe a feeling of calm and relaxation that comes over them as they feed their babies. Brain oxytocin is believed to be part of that as well as causing the "letdown" of milk to enable breastfeeding. It tames the stress system, turning on the chemistry of calm.

But while oxytocin facilitates parent/child bonds—as with prairie vole mating—one or two experiences aren't sufficient to create connection. Despite claims to the contrary, human mothers and infants don't form their primary bond in the moment after birth—bonding takes intensive repetition. In fact, in our hunter-gatherer past—and even in some societies today—infanticide, though rarely discussed, is not uncommon immediately after delivery. Across cultures and time, women have obsessively counted fingers and toes, critically examining new babies the moment they arrive in the world. Today, this practice is usually benign. However, more often than we'd like to think, having the wrong limb numbers or features has doomed babies. Deformed babies, those born during a time of scarcity, and even normal girls in some cultures are discarded and left to die after such examination. Newborn cuteness is designed to trigger empathy to avoid this outcome—but things like birth defects and prematurity can reduce cuteness. If the circumstances of the birth are ill timed or don't meet cultural expectations, cuteness isn't always enough to save the child.

Note however that most infant abandonments occur immediately after birth. They are far less likely to take place after the mother's milk has come in, bringing with it the ongoing flow of oxytocin and the constantly repeated exposure of mother and child that is ultimately responsible for their bond. The cases of later infanticide that do occur are almost always linked with postpartum depression. Abandoning a newborn is always difficult; doing so after a bond has begun to form can be emotionally devastating. As Sue described Galina, it was clear that it wasn't just the daughter who had adoption-related loss issues. "Eugenia's birth mother cried a lot while we were together and when we had to leave," she says. "She has an even bigger hole in her heart, I think."

Indeed, once bonding has occurred, when mothers are away from their babies, they suffer the same "heroin-withdrawal-like" effects seen in separated prairie vole partners. This includes not just anxiety, but, as with separated lovers, obsessiveness. Many new mothers describe bizarre fears about improbable and previously unconsidered catastrophes befalling their infants, even as they recognize that these concerns are irrational. Whenever the baby is out of their sight, they may find themselves drawn back, checking over and over to be sure he is still breathing as he sleeps. These reactions are similar to symptoms of obsessive-compulsive disorder—and women with OCD often find that it gets worse during pregnancy and their babies' first few months, when oxytocin levels are high.

And, as we've seen, oxytocin also seems to be critical in wiring babies' stress systems as they bond with their caregivers. It does this, as Carter once told Maia, by "taking over older parts of the nervous system and putting information into them about a sense of safety and trust." Love via oxytocin relieves stress. Oxytocin is the substance that connects stress relief, relaxation, and calm with the baby's specific caregivers. This, then, is why babies need intensive care from the same small group of people repeatedly and why orphanages like the one in which Eugenia spent her early life can be so damaging. Oxytocin isn't a general "love drug." It's not MDMA (the drug known as ecstasy), though that substance does seem to produce at least some of its well-known empathy-increasing effects by increasing oxytocin release. Instead, oxytocin normally builds connections over time between specific familiar people and stress relief.

Consequently, if infants don't get care from the same few people over and over, oxytocin can't wire their unique attributes to comfort. In orphanages, infant care is done in shifts, and each caregiver spends only a tiny fraction of the time that a parent would spend with her baby. This doesn't provide enough repetition for oxytocin connections to be formed. And if the specific connection isn't made between particular people and trust, the later general positive association between human contact and the pleasure on which it relies simply won't be made. Early experience forms the template for later life—and oxytocin guides its wiring in the case of attachment and bonding.

In fact, a study of children raised in orphanages and later adopted found lowered levels of vasopressin and reduced responsiveness of oxytocin to interactions with their adoptive mothers. In children whose infancies involved normal parenting, oxytocin rose in response to being held by their mothers—but this response wasn't seen in the orphanage children held by their adoptive mothers, even though they were tested approximately three years after adoption.[15]

EUGENIA'S LIFE TODAY continues to reflect her past—even as she's made tremendous progress and made her adoptive parents extremely proud. Sue says that when she first considered adoption, "Many of our friends thought we were nuts: 'Oh my god, how could you adopt these kids, you don't know their gene pool!' Ken said to me this week, 'Sue, imagine if we had documentation on [Eugenia's] gene pool that said her mother was living in a mental institution and the sister's not right.' Well, we didn't know and it didn't matter. She's beautiful!"

By the time Eugenia was twelve, she had told her parents straightforwardly that she wanted to meet her biological mother. Sue has supported Eugenia's quest to get to know Galina, and she and Ken began to arrange a visit. However, because of the vagaries of international adoption law, Eugenia couldn't enter Russia, without risking legal hassles, until she was eighteen. Consequently, Ken and Sue flew Galina to meet them in Romania. By then, Eugenia was thirteen. Around this time, Eugenia learned that she had a severely mentally handicapped older sister—a sister whom Galina had not given up for adoption. Natasha, she discovered, was thirteen years older than Eugenia and lived with Galina in an institution in Russia for the mentally and physically handicapped. This turned out to be where Oksana Malaya, the girl raised in a kennel, now lives. As is the practice there, Galina received room and board for working at the site, which includes a productive farm. Natasha wasn't able to leave Russia for the visit.

"I was real happy at first, but then I got kind of sad," Eugenia says, of meeting her mother and learning about her sister. Sadly, Natasha died of brain cancer before Eugenia was able to meet her. Of course, learning that her mother chose to keep her sister, but not her, can weigh on Eugenia. "Sometimes I hate her," she says of Galina. "Usually I don't. I want to see her again."

Despite these sometimes mixed feelings, her visits with her biological mother help Eugenia feel more grounded and less alone. Just before she met with Maia, Sue and Eugenia had spent five days in Russia. While Eugenia took weeks to recover from the first visit and started to get counseling to help her process it, this time everything went more smoothly. In Galina, Eugenia can see some of the genetic history reflected in her own face and hair. She can speak her first language with her. Sometimes, Eugenia has felt conspicuously different among her cousins or other members of her adoptive family: with the woman she calls her "first mom," she feels like she physically fits in. Curiously, however, she remembers almost nothing about what her mother actually says to her.

"She was very calm, she was very nice, she told me a lot of things and I can't remember any of them," Eugenia says, herself bewildered by her own inability to recall much from such important life events.

Not long after her first visit with her birth mom, in fact, Eugenia's parents had her evaluated because of other memory problems. She seemed to ignore directions and often couldn't recall what people said to her, both at home and at school. They didn't think she was being deliberately disobedient—but the frequency with which she would fail to comply and her classroom memory problems were worrisome. She was diagnosed with an auditory processing disorder: apparently, she wasn't fully taking in what she was hearing.

Eugenia's memory failures—which were exacerbated by stress—occurred because her brain probably wasn't properly decoding what she heard in the first place. Consequently, recall was impossible. Fortunately, she has now developed habits, like using outlines, that prevent the disorder from hurting her academic success. Eugenia sees her auditory disorder as another result of her time in the orphanage. Children who have experienced that kind of neglect are more prone to the problem. Although the cause is not clearly understood, it could result either from overwhelming early stress or from simply having not received focused verbal attention and instruction early in life. So much in the brain requires appropriate exposure at the right stage of development to function optimally—and extreme stress can interfere with this.

Eugenia also sees connections between other symptoms and her early life. Sometimes, for example, she has periods of "numbness" followed by what she describes as "kind of tantrums," where she cries in great distress but does not roll on the floor or physically act out. As a child, she had intense outbursts, too. She says Sue was always able to help her out of these states by "talking to me really softly." All of these symptoms are congruent with common effects of trauma and stress on the developing brain. Eugenia's experience was nearly the opposite of Jeremy's—rather than having her mother try to prevent her from experiencing any stress, once in the orphanage, she was bathed in it, with virtually no buffer at all. Most of the time, when she cried out and needed a familiar touch, she didn't get it.

And so, eventually, she adapted to this lack of nurture by alternating between two states. One was an alarm state, reflected in elevated heart rate, blood pressure, and high activity: at its most extreme, this produced those rage- and sadness-filled tantrums. Eugenia's climbing and hyperactivity when she was first adopted were probably physical manifestations of this physiology. The other state she often experienced was a dull, spacey mood of feeling completely numb, an experience known as a "dissociative state." This comes from the stress response associated with surrender, the one that occurs when the body recognizes that escape is impossible and help won't come. Here, heart rate and blood pressure are lowered, and the person withdraws within, becoming as physically distant and small as possible, shutting out the world. This state is originally a last-ditch attempt by the body to minimize injury—but it can recur later in nontraumatic situations that recall the original trauma in some way.

Adults with trauma-related mental health issues, including posttraumatic stress disorder (PTSD), often share these symptoms: numb and dissociated states, alternating with hypervigilance. In neglected children, however, depression is more common than the classic full-blown

PTSD that might include nightmare or flashback-like memories. And indeed, Eugenia only had nightmares early in life, but she does suffer from depression, which has now successfully been treated.

Her orphanage years have left less obvious scars, too. "I don't like to be touched. Ever," she says. Not being held much early in life has made physical contact unfamiliar, novel, and, therefore, anxiety provoking, rather than pleasurable—it's something she has to control if she is to feel safe. She also has sensory differences. She needs absolute dark to sleep, finds bright lights bothersome, and, as a child, needed to wear only soft fabrics because other clothing was unbearable. Her friendships aren't as close as those she sees between other people. "I think I seem distant," she says. "A lot of my friends hug and are very close, but I don't get that close to people. If I never saw them again, I wouldn't care." That doesn't mean that she doesn't like her friends or treat them well—but she sees that they are connected to others in ways that she doesn't quite understand. She's had to devise cognitive and intellectual strategies—ones that rely on the more plastic, later-developing brain regions of the cortex—to bolster connections that she doesn't always feel emotionally because the social and relational systems in the lower regions of her brain didn't get the affectionate stimulation they needed during an important phase of development. To soothe herself, she sometimes simply sits and rocks.

If those symptoms sound more like signs of autism than the results of developmental trauma, that's because severe early neglect can indeed sometimes mimic autistic conditions. Extreme tantrums, sensory problems, difficulties with aspects of social interactions and empathy, and repetitive, primitive self-soothing behaviors are common to children with both conditions. Although in the past this resulted in tragically misguided parent-blaming—viewing all autistic children as neglected and abused—now the similarities could lead to a better understanding of both, as we'll see in the next chapter.

In the fall of 2009, Eugenia started college. She's planning to study psychology or another area of science that will allow her to explore the brain. She wants to add to the understanding of early neglect, using what she learns to help herself and others like her. While she hopes that orphanages are abolished, if they still exist, she says she'd like to adopt a child from the place where she spent much of her earliest life. She'll continue to build her relationships with Galina and with her family here in the United States. She's looking forward to dorm life. And she's continuing to volunteer with animals. She pets her guinea pigs and says, "They don't judge you and look at you and say anything. If they know you—as long as you treat them well—they don't [hurt you]." She adds, "I guess I control them. Like, when they're in their cage, they can't run away."

four | INTENSE WORLD

MIT MEDIA LAB IS NERD HEAVEN. A large teddy-bear-like robot with an emotive face lingers in the uncanny valley between cute and terrifying. Screens of various sizes, computers, toys, even the odd houseplant clutter most open spaces. Air ducts and other usually hidden building viscera are exposed. Just outside the office of Matthew Goodwin, the lab's director of clinical research, there's an open meeting area with a red British telephone booth occupying one corner. Decommissioned computer mice hang from the ceiling. Nearby, atop a red beam, curls a gray stuffed cat, "an early iteration of Ceiling Cat," someone jokes, referring to one of numerous self-referential and ever-evolving Internet in-jokes.

Although it's clear that the elite researchers here spend plenty of time at play, that openness and willingness to entertain even the strangest ideas serves a serious purpose. Goodwin and colleagues Roz Picard, Rana el Kaliouby, and others are trying to develop technologies to help people on the autism spectrum communicate better. Sam, a software engineer with Asperger's syndrome, a type of autism on the milder end of the spectrum, has met Maia here to connect her with these researchers, tell his story, and explore what people with autism spectrum conditions can teach us about empathy.

Autism is often described as a deficit in empathy, but increasingly, researchers are beginning to discover that this may not be the heart of the condition. Although difficulties with empathy are clearly involved, not all aspects of empathy are equally impaired, and some may actually be enhanced. Recognizing which elements of empathy are affected by autism spectrum conditions and what this tells us about the brain can lead to a greater understanding both of these syndromes and of what empathy is and how to encourage it.

Sam began to suspect that he had Asperger's syndrome when his younger child, Jonah, was diagnosed with another type of autism. At two and a half, Jonah had yet to speak and his play was oddly limited and repetitive. Sam and his wife were becoming increasingly concerned. They'd seen their older daughter's development progress much more predictably. At the beach, Sam remembers building a sandcastle for Jonah. The boy spent hours plopping sand into the water in the moat from his shovel, over and over again, entranced by the motion. The rest of his play was similarly mechanical and isolated. Unlike Sam's daughter, Jonah rarely babbled. He echoed certain words, but that was about it. Nonetheless, the first time they had him evaluated, the clinician thought he was too affectionate to have autism and ruled out the diagnosis.

As Sam soon learned, however, the condition falls on a continuum, ranging from Asperger's syndrome to classic full-blown autism. Jonah's type is called "Pervasive Developmental Disorder—Not Otherwise Specified," which means he has many autistic characteristics but doesn't fit precisely into a particular autism diagnosis. As Sam likes to say, "If you've seen one person on the spectrum, you've seen one person on the spectrum"—the conditions are quite variable. Nonetheless, there are three required diagnostic clusters of symptoms, all of which Jonah has to some extent: difficulties with social interactions like eye contact and, later, making friends; differences in the use of language, including, in the most severe cases, no speech at all; and repetitive behaviors and obsessions—like Jonah's behavior at the beach or repeatedly spinning the wheel of a toy car. Although this aspect of the condition is not included in the official diagnostic criteria, virtually all people with autism and Asperger's syndrome report various sensory problems, usually oversensitivities, but sometimes reduced perceptiveness as well, particularly of pain. Many researchers now believe that it is these sensory problems that directly or indirectly produce the other symptom clusters.

Sam, for example, was so sensitive to sudden, loud noises as a child that other children sometimes terrorized him by intentionally startling him. He also found photographic flashes extremely frightening. By early adolescence he had learned to carefully watch the photographer's shutter finger to give himself the split-second of cognitive "advance notice" he needed to tolerate them. As a child, there were very few foods he would eat. When he realized in early adolescence that most of the gustatory aversions he had were to dairy foods, it became much easier for him to try new (non-dairy) foods. Jonah, too, is sensitive to startling noises and to being unexpectedly touched. As a young child, his avoidance of new tastes led him to limit his diet to little beyond pizza and pasta without tomato sauce.

The more Sam learned about the autism spectrum in the course of researching his son's condition, he says, "the more I realized that Asperger's syndrome as described in the literature might be a more intense version of my own developmental trajectory." Ten months after Jonah was diagnosed, his father was, too.

Sam, who is in his early fifties, has a beard that is starting to gray and a professorial look. He remains slightly socially awkward, but he has trained himself to rein in his digressive and what he calls "syncopated" conversational style when necessary. At MIT, he easily joins the conversation as Goodwin demonstrates an experimental technology meant to help nonverbal or less verbal autistic people communicate with people around them.

The device measures skin conductance—detecting changes in electrical activity on the skin that reflect the activity of the autonomic nervous system. The autonomic nervous system is intimately involved in the stress response, and, mostly, you can't consciously control it. It used to be called the "involuntary" nervous system for that reason. It has two parts—and most of the time, they act in opposition to each other. When what's known as the "sympathetic" part of the autonomic system becomes activated, you experience the typical "fight-or-flight" reaction, with elevated heart rate and blood pressure and energy sent to the muscles, rather than the digestive system.

The "parasympathetic" system is its calming opposite, ordering the body to "rest and digest." Here heart rate and blood pressure go down. You might call the parasympathetic system the empathetic system because it allows the relaxation necessary for social connection and uses the bonding chemical oxytocin as one of its chemical messengers. The parasympathetic system has exactly the links you'd suspect for one involved in easing social contact and helping love conquer fear: it has connections to the heart, brain, face, gut, and genitals. This is another reflection of the fact that most people—at least most of those whom people in the autism world call "neurotypical"—are set up to be wired to find friendly social contact soothing.

Increases in skin conductance reflect increased emotional arousal, which is usually accompanied by elevations in the activity of the sympathetic nervous system. This is why skin conductance is one measure taken by "lie detectors"—it picks up nervousness and anxiety. However, measuring skin conductance can't distinguish whether the arousal comes from a positive or negative emotion. Both joy and anger can increase conductance.

When Sam arrives, Maia is wearing the device. Right now, it consists of two wristbands made of terry cloth athletic sweatbands with electronics inside, one worn on each arm. There are two because changes on different sides of the body can reflect different regions of brain activity. The bands are wirelessly connected to a computer that graphs any changes in skin conductance—with only a tiny delay from real time.

Goodwin tries to pronounce "plesthysmography"—a term for the output of a device that measures blood flow. He's talking about other ways to measure important autonomic information. Maia recalls the only other context in which she's heard the word. "Penile plesthysmography" is used to study sexual arousal, often in sex offenders. As she mentions this, she becomes embarrassed. The line on the screen representing her skin conductance shoots up wildly. When she points this out to the group, she becomes even more self-conscious, blushing and sending her levels even higher.

So why would such a device be useful for autistic people? Maia's chart clearly illustrated a change in her emotional state, one that might have been obvious to someone who was watching her closely but that probably wouldn't have been noticed otherwise. Such changes occur in response to all kinds of stress—and specific stressors can trigger tantrums and other kinds of loss of control over behavior in autistic people. They can become overwhelmed by something they perceive but be unable to communicate the nature of their distress.

This communication problem is most pressing among those who are completely nonverbal. However, even those who speak are not always aware of how stressors affect them over time. Things build up. One distracting noise, light, or touch might not be disastrous, but continued exposure or additional discomforts can synergize, making predicting and preventing outbursts difficult. As the tension mounts, a child's facial expressions or activity might offer little hint of an impending storm.

Alerting a family member, teacher, or the child himself to changes in skin conductance could provide an early warning signal. Familiarity with sensory experiences or situations that have previously caused problems would then allow adjustments to be made before they build up enough to produce an outburst. Since the device is small and portable and readouts could be sent wirelessly, it's very practical. As Sam notes, in normal development, the "terrible twos" are marked by a "temporary excess in the ability to experience frustration compared to the ability to communicate." In typically developing children, as their ability to communicate improves, the tantrums abate. Autistic people, however, may have fewer channels of communication, so a cheap, safe device like this could potentially be tremendously helpful.

It might also help in the care of formerly neglected or abused children—by similarly cuing their caregivers to changes that might be associated with impending tantrums or, alternatively, with "tuning out" into a dissociative state, which would also not be compatible with learning. Children exposed to trauma—like Eugenia—often have lasting changes in their stress response. Some settle into primarily hyperaroused states—like she did as a toddler—others, into primarily dissociative states. Still others swing between these states, typically spurred by even modest stress into one extreme or the other; they have difficulty returning to or finding a calm, normal state. Preliminary research conducted by Goodwin and his colleagues has found that autistic children, too, have a similar range of disturbances in their stress responses and similar difficulties modulating them.

The development of this device represents the increased scientific understanding of the basic nature of autism. It reflects a growing recognition of the importance of emotion, stress, and sensory experience—factors that were often ignored in the past. Says Sam, "Historically, part of how autistic people have been ill-served is that attention hasn't been paid to whether there is some sort of sensory stressor at the most basic level. If we were trying to hold this conversation in the midst of a gaggle of jackhammers, we wouldn't be able to do so: there'd be sensory [overload]. That is really part and parcel of it." For autistic people, ordinary room sounds—or other sensations— can actually be that overwhelming.

Once thought to be rare, autistic spectrum conditions are now known to affect between 1 in 100 and 1 in 150 people. In recent years, autism rates have risen dramatically. Although much of this rise can be attributed to greater awareness and to symptoms that previously might have been seen as anything from odd behavior to mental retardation or even schizophrenia, many experts do believe that autism rates are now genuinely higher than in the past. Various theories have been proposed as to why, but none so far has adequately accounted for it.

Certainly, genetic risk factors are involved, as suggested by the fact that Sam and his son are both on the spectrum. If one identical twin is autistic, the other twin will have an autistic spectrum condition about 60 percent of the time. In fraternal twins—who, unlike identical twins do not share 100 percent of their genes—the odds of the second having an autistic condition are only 0–10 percent. Obviously, then, genes play a big role but are not the whole story. Although environmental influences must account for the remaining risk, few have been scientifically proven to increase autism rates. Behavioral similarities between severely abused and neglected children and autistic children once led psychologists to blame cold, neglectful "refrigerator mothers," but there is no evidence that parenting style has anything to do with autism.

Autism does, however, appear to be related to certain types of skills and interests. For example, college students studying engineering, math, and physics are six times more likely to have a close autistic relative compared with those majoring in literature[1]—and a survey by the British National Autistic Society found that its members were twice as likely to have fathers or grandfathers who work in those kinds of professions than the general population.[2] Sam's family fits into this picture. Sam's grandfather, for example, was exceptionally skilled in math. Sam and his father also mastered written language early—a less known but common experience in Asperger's—teaching themselves to read at ages four and three, respectively. Sam himself was placed in gifted programs as a child. He graduated third in his high school class and attended a prestigious university. Although some cases of autism are associated with low IQ, people on the autistic spectrum are also more likely than those outside it to have what are known as "savant" skills: prodigious talents in music, mathematical calculation, and art. Even some of the most disabled people with autism sometimes develop such gifts.

In fact, some researchers suspect that the same genes that give people an edge in music and math could also carry risk for autism when expressed in certain circumstances. This may be another example where "too much of a good thing" is harmful rather than helpful. It is also a warning sign that trying to reduce autism rates by genetic selection could have harmful effects. Attempting to prevent autism by eliminating these genes could also reduce the benefits and talents associated with them.

Earlier theories of autism—such as the idea that autistic people always avoid physical affection and the notion that they like things but never connect with people—painted a picture of people without emotion, who didn't care about others. But new research, as well as the experiences of people with autistic conditions and their families, suggests that something else is going on. In fact, in some cases of autism, people may actually care too much, not too little.

To understand how this could be true, it's important to note that there are two parts of empathy. We've already explored some of the emotional part: feeling the feelings of others through emotional contagion, like newborns crying in unison. But there's another aspect as well: a cognitive understanding that other people with separate minds are out there and that while you can often share their perspective, they know, feel, and perceive different things, too. Understanding this is called having a "theory of mind." Using your theory of mind is called "perspective taking." (As we shall see in the next chapter, one way that you know that your child has attained theory of mind is that he or she begins to lie.)

A classic series of experiments illustrated the difficulties that autistic children have with theory of mind. Children are shown two puppets, Sally and Anne. Sally has a basket in front of her, while Anne has a box. Sally puts a marble in her basket, then leaves the stage. While she's gone Anne takes the marble out of the basket and hides it in her box. The children are asked: where will Sally look for her marble?

Normal four-year-olds know that because Sally didn't see Anne take the marble away, she won't know where it is now and will look in her basket. Children with mental retardation at ten or eleven—even though their verbal IQ is equivalent to that of a three-year-old—can also understand this. But in the first of these studies, 80 percent of ten- to eleven-year-old children with autism gave the wrong answer, even though the verbal IQ of this group was more than twice as high as for the Down syndrome children the same age.[3] And later studies found a similarly dramatic difference.[4]

So what did the autistic children think? Where did they expect Sally to find the marble? They thought she would look in the box. They knew that Anne put it there. They saw her do it themselves: why wouldn't Sally get that? In other words, they didn't realize that other people aren't aware of the same things that they are. They didn't understand that other minds have different perspectives.

For most people, theory of mind is so intuitive that it's hard to imagine not having it. This ability is required to be able to do anything social at all. However, understanding that autistic people have problems with this helps explain a lot of their behavior. For example, if you thought that everyone else already knows what you know, why would you talk much?

Problems with perspective taking also help explain the refreshing honesty of many people with autistic spectrum conditions. It makes no sense to lie if you think others already know the truth. The level of difficulty this presents in terms of ordinary socializing is hard to imagine. For example, people with autism spectrum conditions often don't understand things like why you shouldn't respond "Yes, you do look fat in that!" when a friend or partner asks about whether an outfit is flattering. They take the question literally, without thinking that what the other person really wants is social reassurance. And they do this because, not realizing that others see the world differently, they can't put themselves in their place and project what they'd want to hear from there. It's this part of empathy that causes problems for them.

Sam points out that people with Asperger's may also have problems with empathy because their minds are so different from those of other people. For example, a ten-year-old boy might be fascinated by the periodic table or prime numbers, discussing his obsession endlessly. Unfortunately, this usually doesn't make him popular with other children. Indeed, children with Asperger's are often known as "little professors" because of the way their intellectual interests dominate their lives to the exclusion of everything else. Such children clearly do understand that other people have separate minds—but, like everyone else, when they take other perspectives, they project their own point of view onto them. The boy obsessed with prime numbers won't stop discussing them because he thinks that if they fascinate him, they'll fascinate you, too. He doesn't recognize the cues others send of boredom—or the need to take turns. But that doesn't mean he's unfeeling or uncaring. You could call this "mis-empathy"—it's not a failure to "mind read," just a misreading.

Sam notes that people who aren't autistic are also "rather lousy at understanding inner state of minds too different from their own—but the nonautistic majority gets a free pass because if they assume that the other person's mind works like their own, they have a much better chance of being right." On the odds alone, if your mind works similarly to those of 99 percent of others, your guess about how you would feel in someone else's place is more likely to be accurate than if you are in that last 1 percent and project your sense of their position onto another.

And indeed, this idea is backed by research on what is known as "empathic accuracy." One study looked at patients with a syndrome completely unlike those on the autism spectrum: borderline personality disorder (BPD). Sufferers tend to have explosive love/hate relationships and an inability to deal with rejection. (Glenn Close's character in the film *Fatal Attraction*—the scorned mistress who boils the pet rabbit of her lover's child—is an extreme, negative example.) Therapists had long suspected that BPD made people more sensitive to other people's emotions (though not necessarily more likely to respond empathetically to them). In the study, researchers had patients with the disorder interact with people who did not have it—and indeed, the people with BPD read the others better than the unaffected people read them.

When this was examined more closely, however, the differences were found to be related to the failure of the people without the disorder to predict the behavior of the people with BPD. In other words, the borderline patients seemed better at reading other people only because other people were so bad at reading them![5] If your mind works differently than others', it is harder for them to stand in your shoes.

The same, of course, is likely to be true for unaffected people trying to understand people with autism.

Difficulties with the cognitive part of empathy also play a role in many other conditions and experiences. For example, depressed people often perceive others around them as hostile or negative, even when they are not. If you walk into a party assuming that everyone dislikes you, it's quite easy to rapidly find confirmation of your idea—you may interpret innocent looks as "dirty" ones and mistake neutral gestures for rejections.

Some studies find that people with depression actually fail to perceive smiles that are directed at them. It's easy to see how these kinds of distortions in perception and perspective taking could enhance or prolong depression or social anxiety. Unfortunately, even just feeling lonely can, by itself, increase these misinterpretations.[6] And sadly, if—as seems to be happening today—more people become depressed, disconnected, and lonely, these mistakes in cognitive empathy could have a self-reinforcing effect, making society even more alienating and contact even more difficult to make. The more people misinterpret one another's positive or neutral signals as negative, the harder it becomes to connect.

Sam himself has had a long struggle with depression. But being diagnosed with Asperger's helped alleviate some of his distress by allowing him to understand some of the barriers that he has faced. It's clear from speaking with him, as well, that he is deeply emotionally empathetic, although this is sometimes expressed in ways that aren't immediately apparent to those unfamiliar with him.

In one interview, Maia asked about his experience of emotional connection to others. He replied in a way that at first seemed completely unrelated to the question, explaining how particular pieces of classical music served as touch points for his key emotional memories. His tone of voice seemed matter-of-fact. When he mentioned Brahms's

German Requiem, however, his voice broke. He said that it had been one of his mother's favorites. Even though she'd died twenty years earlier, just talking about listening to it made his voice catch. "I think that it's a stereotype or a misconception that folks on spectrum lack empathy," he says, explaining that, as he'd just done, they may express emotions differently. "I think most people with autistic spectrum conditions feel emotional empathy and care about the welfare of others very deeply."

He adds, "There are things that have choked me up in unusual places and I have had to struggle to keep my voice from breaking and revealing an intensity of emotion at that point anywhere from distracting to embarrassing." In an online community of people on the spectrum called Wrong Planet, numerous posters also agreed with the idea that many autistic folks struggle with a surfeit of empathy. "If anything, I struggle with having too much empathy," one person commented. "If someone else is upset, I am upset. There were times during school when other people were misbehaving, and if the teacher scolded them, I felt like they were scolding me."

Said another, "I am clueless when it comes to reading subtle cues, but I am *very* empathic. I can walk into a room and feel what everyone is feeling, and I think this is actually quite common in AS/autism. The problem is that it all comes in faster than I can process it."

As a result of that overload—just as with the kindergartners who became less empathetic when other people's pain caused them too much distress and the young nurses who avoided the dying because their empathy toward them was too upsetting—some autistic people can end up behaving less empathically because they feel more so.

BUT IF LACK of empathy isn't what defines autism, what does? And why would children like Eugenia who suffered early neglect exhibit similar

kinds of behavior? A new theory of autism, devised by Swiss researchers Kamila and Henry Markram, could answer both questions, putting all the disparate symptoms together and offering further clues
about how the brain develops empathy. Henry's son from a previous
relationship has a high functioning form of autism—and the couple's
observations of his overwhelming experiences of fear led them to for-
mulate what they call the "intense world" theory.

The Markrams and their colleague Tania Rinaldi backed up this
theory with detailed studies of rats exposed as fetuses to the anti-
epileptic drug valproate (also called valproic acid, brand name:
Depakote)—a drug that can cause autism in humans exposed in utero.
These "VPA rats" have symptoms reminiscent of autism: they avoid
other rats, are more fearful overall, show repetitive behaviors, and are
oversensitive to things like noises and light. They also have a pattern
of brain development that looks like it would produce that combination of
symptoms—plus enhanced abilities, possibly similar to those seen in
human savants. The theory raises the possibility that all autistic people have the potential for savantlike skills.

When researchers examined the brains of VPA rats closely, they
found that they have more groups of cells that serve as "processing
units" in their cortex. Recall that this is the most sophisticated and
most recently evolved brain area. It includes areas involved in interpreting complex information from the senses. VPA rats not only have
more cortical processing units, these units are also better connected
to one another, at least locally. Together, these characteristics could
intensify both sensory experience and, potentially, intelligence, just
like adding both memory and higher speed connections to a computer
improves performance. However, the rats' long-distance connections
from one brain area to another are not similarly enhanced.[7]

Interestingly, autopsies of both scientists and autistic people have
found evidence for a similar pattern of increased brain tissue and

connectivity—even though the scientists studied were not autistic.[8] Autistic children have also long been known to have accelerated brain growth early in life. At age two, their brains are roughly 5 percent larger than normal.[9]

So why would this be a problem and why would it particularly affect social connections and empathy? Unfortunately, the VPA rats don't just have amped-up local processing in their cortex. They also learn fear-related associations more quickly—and have a harder time learning that a once scary situation is now safe. This is linked to increased connectivity in the amygdala, a lower brain region that detects and deals with danger, among other things. Consequently, not only do these rats perceive more and process more in certain brain areas—they also feel more fear and remember it better. Further, they don't have added neurons for cross-brain connectivity to help them put all of that together. It's easy to imagine why this combination might result in both genius—and, without other compensations— total brain overload.

The scientists whose brains were studied at autopsy were somehow able to manage their brain's increased connectivity without becoming autistic. Indeed, they succeeded at high levels and did not appear to have social problems. Consequently, there must be some situations in which this is compatible with normal development, although right now no one knows what makes the difference. But in autism, like a fast computer with a slow Internet connection and slow internal long-distance wiring, the power of the brain to communicate both long distance within itself and with the outside world could be compromised. Add too much sensory information to intensified fear, and you, too, might want to sit in a safe place and rock. Repetitive behavior may be both an attempt to soothe the stress that comes from not being able to take comfort in relationships and a way to manage the overload and reduce overexposure to this overwhelming novelty.

If this is what's going on, the social deficits seen in people with autism could be secondary to the heightened fear and sensory issues, and not necessarily a sign of inherent problems with the social regions of the brain. Autistic children might develop repetitive behaviors for the same reasons that neglected children do: as a way of coping with stress. As another way to manage sensory overload, autistic children might increasingly withdraw within. Social stimuli are among the most complicated we face. If you can't handle loud noises and sudden activity, imagine what preschool must be like! However, if you hide away and avoid social experience, your brain isn't going to have the opportunity to learn from it.

This would mean that something similar to the cause of social problems in neglected children is responsible for those problems in autistic children: extreme patterns of stress response system activation and a lack of appropriate stimulation at the right time. Being constantly bombarded with too much information in itself is stressful; autistic children might act like traumatized children because for them, normal experiences can be so overwhelming as to be traumatic. It doesn't matter whether your brain is missing necessary stimulation because you are in a barren orphanage or because you avoid physical contact and loud, unpredictable peers due to sensory overload. Either way, your brain isn't receiving what it needs for optimal development.

The same lack of relevant stimulation could be responsible for problems with language development in autism. Bruce has seen cases of children raised in cages—like Justin,* who was raised with dogs and gave our last book its title, and Oksana Malaya, the little girl raised in a kennel who was a friend of Eugenia's sister. If they aren't reached early enough, their full ability to use language can be impaired for a lifetime. Some—usually those rescued later than puberty—may never speak full sentences.

Even a capacity as basic as seeing requires visual information to reach the brain at the right time in order to develop properly. For example, cats with perfectly normal eyes can go blind if they cannot open their eyes during a specific time period when they are kittens. If this happens earlier or later, there is no damage, but if their brains don't get visual information during this "sensitive period," they will be blind for life. Similarly, language exposure is necessary early in life for the full development of fluency and grammar. Evidence for this is seen in babies' incredible ability to learn two or even three languages with little effort and no "foreign accent" if they hear them used regularly in daily life by familiar people. On the other hand, trying to learn a new language after puberty requires serious, laborious effort—you'll almost always have an accent even if you become fluent. That's because the most highly plastic, sensitive period for development of the brain's language systems has passed.

Now imagine the world from an autistic child's perspective. Frequently, he is flooded with overwhelming amounts of information—but he has little ability to express the problem. Alternatively, he may experience periods in which little or no information gets through. It's like he's at the front row of a rock concert near a blaring speaker trying to figure out what someone next to him is whispering—and then in total silence with no sound coming through at all. It's impossible to filter out the unnecessary information or home in on the important stuff. Not surprisingly, he retreats into repetitive behaviors and craves routine: at least predictability can minimize the fear that results from this chaos.

But this retreat has serious consequences. While Jonah was sitting on the beach dropping sand into a bucket, other children the same age were interacting with their parents and one another, over and over and over. They were learning not only verbal language, but body language and other implicit social information. While the autistic child is overwhelmed by sensation or getting only minimal input, other children are hearing people speak to them; when the autistic child is off in a corner rocking, he is missing many of the repetitions that children need to get to learn language and understand that other people have different perspectives, thoughts, and feelings.

Researchers are now studying whether children with the most devastating experiences of sensory overload or fluctuation also have the most severe autism. If this is true, the reason that some autistic children don't speak could be that because of their sensory problems, they are simply not getting enough exposure to language. If, as a baby, all you could hear of a conversation were random bursts on a poor cell-phone connection or badly tuned radio, you probably wouldn't be able to learn to talk, either. In fact, you might not be able to understand that people were trying to communicate with you at all. And if this weren't discovered early enough, your exposure to language might be—from your brain's perspective—as limited as that of Oksana, living with dogs, not humans. If that was the case, fully developing empathy or even learning to speak would be close to impossible. But not because you didn't have the inherent capacity.

Similarly, deaf children whose disability is discovered late can appear to be mentally retarded and can have lifelong learning disabilities—again, not because their brains don't have the ability to learn but because they didn't receive necessary early stimulation. Since infants are now routinely screened for deafness, late diagnosis has become rare and most are exposed to sign language right away. As a result, many cases of mental disability that were once thought to be inherent to some types of deafness have been eliminated. Things that look "fixed," like mental retardation linked with deafness or social withdrawal in autism, may really be caused by environmental deficiencies, not lack of genetic capacity. In fact, a recent randomized controlled study of

eighteen- to thirty-month-old toddlers with autism found that inter-
vening early—with fun, child-directed activities—dramatically improved both IQ and social functioning. The results were so strong that
seven of the twenty-four children given the intervention (known as the
Early Start Denver Model) did not receive an autism diagnosis when
evaluated by psychologists who did not know whether they'd been
treated or not—compared to just one child out of twenty-one in the
control group.[10]

Unfortunately, if a child misses that experience during those sensitive periods—which seem to occur during the first three to five years of
life—many, many more repetitions are needed than if stimulation had
been accessible at the right time. The brain remains "plastic" and able
to change and learn throughout life, but being under- or overstimulated
during sensitive periods, particularly for language and emotional development, can reduce its abilities profoundly. This can create conditions
that appear "genetic" and unchangeable. Bruce's work with severely
neglected children has shown that giving them the responsive, rhythmic, and repetitive experiences that they missed can help even some of
the most severely affected catch up. But sadly, most of the time, neither
schools, nor parents, nor society have the resources or persistent pa-
tience necessary to provide the truly intensive care needed to do this.

The intense world theory suggests that autistic children may be
failing to receive necessary social input during this critical period—
because of sensory issues, not because of problems inherent in their
social brains. If it is correct, it offers a great deal of hope to families
with autistic children. Catching sensory problems and mitigating
them early might allow autistic talents and passions to develop—
possibly without social or empathy problems developing at all. Just as
placing Eugenia in an adoptive home without any time in an orphanage would probably have eliminated her auditory processing problems
and dramatically reduced her sense of feeling distant from others, re-

ducing sensory problems in autism could potentially allow these children to learn language and socialize normally.

It's likely that the processing style that makes their brains unusual would still have effects—differences in intelligence, intense interests, or other issues are possible—but disabling difficulties with language and reading people could be possibly reduced or even eliminated. Additional savantlike talents could be discovered. Researchers are currently working on interventions that will help make sensory issues less overwhelming—for example, headphones to reduce noise, and visual and tactile environments that are calmer, simpler, and quieter. Other tactics would have to be devised for children whose problems are due to fluctuations and undersensitivities. Researchers at Johns Hopkins are currently testing ways of reducing oversensitivity in children who receive the earliest diagnoses.

Sam has seen how focusing on these issues has paid off for his son and others in the autism community. "You get much further if you can work to mitigate the source of sensory distress," he says. "For that, you need expressive communication, whether via speech or otherwise. If a child doesn't start to acquire expressive speech at eighteen to thirty months, you really need to be working on developing alternative reliable, trusted, and respected means of expressive communication—picture books, keyboards, AAC [alternative and augmentative communication] devices, whatever. If you go that route, you are going to bypass a boatload of behavioral horror stories."

For Jonah, Sam and his wife found a special preschool program to bolster his acquisition of speech. The program worked in part by utilizing—rather than ignoring or suppressing—the way he tended to echo and enjoy repeating words at age three to four. Later they ensured that he was in appropriate schools and continually sought to understand what his needs were, what situations were troubling him, and how to help.

Now eighteen, Jonah attends a high school program for teens with autism spectrum conditions but takes regular education math classes at a public high school. He's reading young adult books and recently took a class in computer repair with his father at a local vocational school. He was heading off to a brief summer internship at a relative's business when Maia visited with Sam. Undoubtedly, having an understanding family devoted to finding the best ways of enhancing Jonah's abilities has made a huge difference for him.

AT MIT, GOODWIN and his colleagues are working not only on devices to help others read autistic people better, but also on those that can help autistic people read others better. El Kaliouby and a graduate student demonstrate a contraption that they have designed to read both facial expressions and gestures. It's called the "interactive Social-Emotional Toolkit" (iSET). Both the iSET and the skin conductance device are being commercially developed by a company that some of the MIT researchers have set up. A video camera locks on to Maia's face. On a computer screen, a series of dots surround her eyes and mouth, following their changes. It notes when she nods her head or smiles, attempting to read her expressions. Ultimately, it will display information like whether she's agreeing or disagreeing. El Kaliouby, who splits her time between MIT and Cairo, first became interested in studying how to read the face when she was away from her loved ones and realized how little emotion is communicated through text online.

El Kaliouby previously worked with leading autism researcher Simon Baron-Cohen (indeed, cousin of "Borat" Sascha Baron-Cohen), a professor of developmental psychopathology at Cambridge University. He's the one who developed the "Sally/Anne" test of theory of mind and did the study finding more engineering and math professionals in families with children with autism. More recently, he developed an interactive DVD in which actors demonstrate facial expressions of particular emotions to help autistic people learn to read and respond to them. El Kaliouby will soon begin a clinical trial, comparing whether the iSET or Baron-Cohen's DVD produces greater improvement in social skills among autistic children. Ultimately, the MIT team hopes to create something small enough to fit inside a normal pair of glasses, which could give the wearer information about other people's emotional states without anyone else being aware that the person is getting this assistance.

Says Goodwin, "The ideal scenario is that you create a technology that empowers people to learn what the technology is teaching them, and then you phase the technology out, and they've got that skill on their own." The team recognizes, however, that not everyone will be able to function without continued use of the machine. Goodwin gives an example of how such a device might help autistic people in the real world.

"Say I have Asperger's syndrome and I am supertalented at programming computers," he says. "I also happen to really really like the Red Sox. The World Series is two days away and they're in the playoffs and I'm having my last interview for a job. And I'm going on and on and on about the performance of this baseball team and the interviewer is wanting me to talk about my problem sets." The interviewer begins showing signs of boredom, looking away and wondering whether the programmer will be able to work with others. But while the programmer can't read the interviewer's signals, the new technology can. It either displays text readable only by the wearer of the glasses ("Now might be a good time to stop and ask a question"), vibrates his cell phone, or in some other unobtrusive way tells the programmer that he needs to end the digression. A situation that might once have resulted in a lost opportunity now becomes much more manageable, as the device prompts perspective taking that might not otherwise have occurred.

At the moment, it is far from being able to work seamlessly. However, already autistic children are enjoying using prototypes to try to read each other and other people. One boy told the researchers his arm was starting to hurt from carrying the prototype—but he still declined to put it down because he was so fascinated by what it could show him about others.

AUTISM SPECTRUM CONDITIONS give us a window into social development and the many skills needed to empathize. The intense world theory, the new emphasis on understanding emotional and sensory aspects of the disorder, and these new technologies could offer hope to affected people and their families. Online, autistic people are also organizing and connecting to one another, fighting the stigma associated with the condition and turning it into a form of "autistic pride," as other disabled and marginalized people before them have done. These connections themselves undermine the notion that autistic people lack the capacity for empathy.

Sam notes that autistic people have often been victims of the mainstream's lack of empathy—rather than failing themselves at connection. And the list of their contributions to music, math, computing, and other fields is long and growing. Baron-Cohen has suggested that both Einstein and Newton showed many of the characteristics associated with Asperger's syndrome: obsession with complex systems, social awkwardness, and in Einstein's case, late development of language. Other Nobel laureates such as physicist Paul Dirac and chemist Irène Joliot-Curie (daughter of Nobelists Marie and Pierre Curie) also shared many autistic characteristics,[11] as do some of the pioneers of computer science who helped create and maintain the Internet.

If it turns out that autistic development is going awry not because the social regions of the brain are damaged, but because those with autism lack necessary exposure to social stimuli, their stories have extra relevance to everyone, no matter where they lie on the autistic or neurotypical spectrum. So, too, do the stories of the effects of neglect. The social brain needs social experience to function; like a muscle, it won't grow if you don't use it. And for many of us today, that muscle is getting weakened.

Children today spend half as much time playing freely outdoors than they did in the early 1980s[12] and significantly more time in structured activities. Free time is necessary to social development: the social brain doesn't grow without practice. On average, children under six spend two hours a day in front of a computer screen—three times as much as they spend reading offline or being read to.[13] Nearly half of babies use screen media every day.[14] We're not saying that the rise in autism spectrum conditions is linked wholly or even in part to these factors—but we would like you to think about how all of these things interact to provide less time for all children to learn empathy, less time for them to ease one another's stress, and less emotional connectedness overall.

The ability to empathize and truly bond with another isn't an "on/off" phenomenon; it is complex and multilayered, and, like autism, it lies on a spectrum. If we shift the empathy spectrum further and further away from its fullest potential, we risk making the world an increasingly less caring place.

five | LIES AND CONSEQUENCES

A S A CHILD PSYCHIATRIST, BRUCE HAS spent a lot of time coloring with his younger patients and interview subjects. It's relaxing and absorbing for them—and often, for him, too. It helps build rapport, giving him a crucial opportunity to connect with the child. Particularly for children who have difficulty expressing themselves in words, their artwork can reveal what's going on in their minds, even if they are afraid to verbalize their fears or concerns. But this sandy-haired, green-eyed six-year-old boy named Danny* confounded Bruce.

Bruce had watched him drawing, creating a picture that might qualify as a work from the "blue period," if he'd been Picasso. "Do you like to use blue?" he asked, hoping he wasn't overstating the obvious. Children aren't stupid—often, even the youngest can sense your discomfort or clumsiness.

"No," Danny replied, his voice emotionless.

"Well, you used a lot of blue in that picture," Bruce said, again, trying to engage him.

"No, I didn't," Danny responded, with a guileless tone.

"Really? What color is that?"

"Blue."

"Well, you drew that."

"No, I didn't." He was matter-of-fact, cool. Bruce had him on a heart rate monitor, which, like the device the MIT researchers use to measure skin conductance, can offer insight into the responsiveness of a child's stress response system. Like skin conductance, heart rate is monitored by "lie detectors" that attempt to determine whether someone is telling the truth. Most people's heart rates will spike when they are worried about being caught prevaricating. Danny's didn't budge.

This was interesting, Bruce thought. He wondered, *Is this boy mentally retarded? Does he have some kind of brain damage or sensory processing disorder?* But there was no other evidence for this, based on previous observations made by Bruce and others. He seemed cognitively normal. There were no signs even to suggest brain damage or sensory issues, which could, for example, interfere with his ability to label colors. The two chatted some more, about nothing in particular, as Bruce continued to try to reach him.

Bruce had been asked by the FBI to interview Danny. As an expert in interviewing children, he is often called upon by law enforcement to help with difficult cases. He was meeting with Danny and his brothers, who had what might be seen as the opposite problem with empathy to that seen in autistic children. They were excellent at perspective taking, but lacking in the emotional, caring core needed for compassion.

Danny drew a few more pictures. Then, getting bored, he got up and wandered around the room. He was in a Children's Assessment Center in rural Indiana, in a room typically used by Child Protective Services (CPS) for interviewing suspected victims of child abuse. The walls were covered in cheery wallpaper in primary colors. Low shelves were filled with slightly battered toys. But just like in a children's cancer ward, these attempts seemed only to underline the unhappiness and suffering that was often expressed here. Among the toys was a tiny doll family, a father, mother, and children. In this room, they were often used to help the youngest victims tell their stories. (These were not naked or anatomically correct: research has found that use of those kinds of dolls can lead to false reports of sexual abuse.)

Although there was some concern that Danny had been abused, this wasn't the reason for Bruce's visit. This was a forensic interview—Danny was believed to have information about a crime. Bruce looked on as Danny went up to the shelves, took a child doll that was about two inches tall and placed it in his pocket. He saw Bruce watching and came back and returned to his coloring.

"Do you like those dolls?"

"Not really," he shrugged.

"You put one in your pocket."

"No, I didn't."

There it was again: a blatant, apparently purposeless lie. And there was no blushing, no wriggling, no movement on the heart rate monitor or visible discomfort. Now Bruce began to understand why the FBI had wanted him to interview Danny and his three brothers. They came from a family of "Irish Travelers"—a secretive group of itinerant people of Irish and Scotch-Irish origin, who had immigrated to America over a century ago from the British Isles.

Some—but by no means all—Traveler groups are essentially organized crime families. The FBI was investigating this particular group because of its suspected involvement in a murder. But most of the crimes the group committed were nonviolent, typically scams and cons targeting the elderly. These included hard sells of unnecessary household repairs like roofing jobs and driveway sealing. If the promised work was done at all, it was shoddy and more payment was demanded to fix problems. Alternatively, payment was simply stolen and no work was done. Danny's family and the rest of the group would come into a community and settle for a few months, often after a natural disaster. They'd pitch themselves as roofers or other construction workers, there to help out. They'd leave suddenly when the locals began to catch on to the scams.

This particular group was made up of several families who'd widely intermarried and a number of occasional members who would join up with them for a while and then leave. They owned various properties that served as a home base. These were located in rural, isolated areas that could accommodate trailers for multiple families. Sometimes, the women would stay behind while the men traveled to "work" for months at a time. Often, the whole family traveled.

Children dropped out of school early, sometimes without starting— let alone finishing—high school. They were on the road at least six months a year. Girls tended to marry extremely young—at the ages of thirteen, fourteen, occasionally, even twelve. In some states, this was legal, with parental consent, which they naturally had. In fact, the parents usually arranged the marriages, at least in part to cement alliances and gain social advantage within the group. Locals were often shocked by the overtly sexualized way that the young girls were dressed.

Danny and his brothers were of interest to the FBI because they were believed to have information relevant to a murder, for which one of their relatives was a suspect. One of their cousins, a trucker, was thought to have killed a young woman who had briefly lived with them. Like many of the people who joined sporadically, the victim had apparently been a "throwaway," a teenager who had been kicked out of her home by her parents when she became rebellious and unmanageable. Another such young woman, Shirley,* had tipped off the FBI to the group's involvement in the crime. Shirley was aiding the investigation after becoming fed up with how she had been treated by a male member. Her earlier call to CPS yielded an investigation that went nowhere.

Workers from CPS, FBI agents, and now Bruce all quickly confronted the same problem. The boys' relationship with the truth was distant, to say the least. Bruce is an expert in conducting these types of interviews, but so far, he was proving to be as inept in getting the real story out of them as everyone else.

HONESTY IS A cornerstone of empathetic relationships. If people aren't willing to share their true thoughts and feelings, forming a connection with them is impossible. Needless to say, trust simply cannot be maintained. As Bruce eventually pieced together, these four boys had been raised in a situation in which they had literally been trained in deception. Interestingly, learning to lie is an important milestone in our development of empathy—but, of course, normal parents quite rightly discourage it, rather than teach it.

Here's why, when a child tells his or her first lie, it's actually a positive developmental step. As we've seen, children are born with the ability to imitate others, which allows them to get a sense of what someone else is feeling by internally copying it. But over time, children realize that they have a unique perspective. When someone else gets hurt, they don't feel physical pain like they do when they themselves are injured. Young children may find listening to another child's cries upsetting— but they learn that what is happening to the other person isn't happening to them. At around age three, children make a great cognitive leap—they begin to understand that while "I" might know and feel something, other people quite possibly might know and feel differently.

And this is where lying comes in. Once a child is capable of realizing that someone only knows about what he's personally been exposed to, he soon discovers that if he does something that Mom and Dad don't know about, they can't punish him for it. Of course, at first children aren't very good at lying. A friend of Maia's reported a prime example. Her nanny suspected that her two-and-half-year-old son had soiled his diaper. The boy vehemently denied it. The nanny checked, and, suspicion confirmed, asked, "Why did you say there was no poopy?" The little boy replied completely ingenuously, "Oh, I don't know who did that," and smiled.

Once a child reaches this stage, he knows that others think different thoughts and see the world through different eyes—but also that these perspectives can often be predicted by imagining "what it would be like for me." Learning to lie helps you consider things from someone else's point of view—failures and mistakes help refine these considerations. If you can't do this at all, you can't empathize, but you also can't lie effectively. This is yet another example of how the positive parts of human nature rely on the negative ones in important ways. As frustrating as they can be, children's lies are signs that they have developed theory of mind and understand perspective taking. In autism spectrum conditions, this development is delayed.

Danny clearly didn't have that kind of problem. He was certainly adept at lying and denying the truth. He did lie blatantly about things like his coloring and hiding the doll that were easy for Bruce to disprove—but he'd also developed emotional control that made it difficult for anyone to discern the truth when it wasn't clearly apparent. In fact, his early experience had produced what might be seen as the opposite problem to the difficulties with lying seen in autism. In his family, there was little trust and, therefore, little motivation to tell the truth unless it could be used to manipulate someone. His parents certainly hadn't emphasized honesty, nor had they focused on sparing people's feelings with white lies. Instead, he was taught that lying was the best way to get what you wanted.

Much of this teaching was implicit: his parents and other relatives frequently and sometimes blatantly lied to him about what they were doing, just as he'd done with Bruce. If they followed the patterns seen in other members of this group—and there was no reason to believe that they didn't—Danny often saw them lie to one another as well. Discipline would have been harsh and physical and applied inconsistently. If a lie could achieve your goal, it was valorized—the ends justified the means. You told whatever story would help you avoid punishment or exposure.

However, Danny *was* clearly taught explicitly to fear outsiders and to keep quiet about what was really going on if they questioned him. The police and other authorities weren't seen as good guys who could help them—they were viewed as enemies to be avoided and deceived if at all possible. And the perspective of law enforcement as a threat had absolutely been made clear to even the youngest children.

Bruce made a few additional attempts to get Danny to open up. He tried talking about his family, first asking how many siblings he had.

"None."

"I thought you had three brothers?" Bruce said, trying to be playful about it.

"Oh, I guess I do." This was one of the few times he got him to back down. But it was ridiculous: his brothers were in the waiting room so it was patently obvious. He was good, though: if Bruce hadn't heard his other lies, it would have been easy to find him credible. And his physiology would have fooled even the most expert interpreters of a "lie detector" test—the existence of people like Danny is one reason these tests are not admissible in court.

By this point, the interview had taken on the characteristics of absurdist theater. It was wordplay, not any kind of meaningful human interaction. Bruce thanked Danny and told him he could go. As he left, Bruce tried to clear his mind, wanting to start the next interview as fresh as possible. He looked over the information that the FBI had provided and saw that his next interviewee would be John,* age eleven.

Bruce hoped he would do better with him. They settled in, and

Bruce sat down at his level to make him feel comfortable, so he wasn't towering over him. But rapidly, he began to encounter the exact same problems. Although a few of John's lies were as obvious as Danny's, it was impossible to tell if his other claims were true or false because they were not accompanied by the normal cues that suggest deception.

By the time Bruce got to the oldest boy, Robby,* who was a teenager, he had begun to think the entire process was futile. Out of idle curiosity more than anything else, he said to the boy, "Let me ask you a question." He described the incident in which Danny had lied about taking the doll.

Robby smiled slightly. "Yeah, he's a liar," he said.

"Where does that come from? Does he lie to you?"

"Well, sometimes."

"Do you ever lie?"

"No, not really."

"So if I ask you things about your life, you're not going to lie to me?"

"No, I won't lie."

But when Bruce began the questioning, Robby immediately began saying things that were clearly untrue. There was a swimming pool at his home (it was a trailer without a backyard). The sky was yellow. Bruce had to tell the FBI that there was absolutely no way that any information could be obtained from these children that would hold up in court—and that interviewing them even for background would be useless. Trying to figure out what was true and what was not would be virtually impossible and probably more time-consuming than any information gained would be worth.

BRUCE HAS THOUGHT a lot about those boys and their lives over the years. He's worked with children from many tightly knit and dysfunctional groups—from extended families with multigenerational histories of sexual abuse to cults like the Branch Davidians. Trust and consistency are clearly important for the development of empathy and morality in the brain. But all too often, the child-rearing practices and beliefs of a group exploit our reliance on one another for stress relief, allowing the leader to take advantage of the most vulnerable, often women and children.

As discussed earlier, the brain develops in a social context: one cannot develop a sense of self without a sense of the other. Humans evolved to be especially sensitive to social cues. In fact, the complexity of dealing with group social life is believed to be one of the key reasons our brains became so big. But like so many characteristics of the human brain, this sensitivity is a double-edged sword. It allows us to adapt to varying social conditions—but the adaptations that are helpful in one set of circumstances can be outright harmful in another. *The relational capabilities that give us our great strengths as a species in the natural world also underlie our major vulnerabilities in our "cultural" worlds.* The further away we are from the tight-knit, relationally enriched hunter-gatherer structure of our ancestors, the more complex our societies become, the easier it can become for certain unscrupulous predators to exploit, manipulate, trick, and dehumanize others.

Let's look more closely at what the brains of Danny and his brothers would have learned during their early life and how these influences can shape behavior. We know that our mirror neurons predispose us to imitate the actions of those around us. Mom smiles, baby smiles; baby giggles, Mom laughs. If things are going well, that's how we learn to share one another's elation and grief and everything in between. For centuries, it was a mystery as to how the capacity to care could have evolved. The discovery of mirror neurons helped explain the mechanism that allows empathy. Combine this with the recognition that natural selection can sometimes favor altruism, and a coherent explanation begins to emerge. In fact, understanding how altruism evolved can provide essential insights into how empathy works—and what's necessary to sustain and nurture it. Interestingly, the roots of altruism—like those of so many positive human qualities—are full of paradoxes, too. So how could unselfishness develop in a fiercely competitive natural world?

Since the days of Darwin, this has been a challenging question. He recognized it as a threat to the theory of evolution itself. Common sense suggests that under virtually all circumstances, selfishness should win. If you always share your food, you'll be less well nourished than those who hoard or steal it. If you let others take the risks during the hunt, you will probably benefit from their effort without endangering your own skin. If you leave your children with your friend but are always busy when she needs help with hers, you have more time to make more babies of your own.

Altruism—engaging in acts that help others but have costs to you—can make evolutionary sense if it is directed toward close genetic relatives, of course. Relatives share many of the same genes. Successfully passing on genes is what drives evolution, so anything that helps families survive and reproduce is likely to be passed through the generations. But altruism toward nonrelatives is less easy to understand. If selfish creatures can take advantage of altruists at no cost to themselves, the altruistic genes should not be favored for survival and should soon die out as a result.

For years, people argued that this meant that true altruism did not exist in nature. They claimed that even though it looked like some animals were sacrificing themselves for others, this happened only with kin or only with some sort of hidden benefit to self that actually outweighed the sacrifice to the other. Some even claimed that the very existence of altruism was proof that Darwin was wrong. The "selfish gene" should always win in the end—otherwise, how could the process truly be driven by individual competition?

Then, in 1973, British biologist John Maynard Smith and American geneticist George R. Price introduced some ideas from a branch of mathematics called "game theory" into evolutionary thought. These concepts involve understanding a situation called "the prisoner's dilemma," which was originally devised by mathematicians Merrill Flood and Melvin Dresher. It goes like this. Two partners in crime (or, if you like, since, in this game, defeating the police is the goal, two political prisoners) are being interrogated. They cannot communicate with each other. The first one who "snitches" will go free; in this case, the other guy will get a ten-year sentence. But if both parties decide to "rat each other out" simultaneously, they'll each serve five years. If both stay quiet, however, each will only serve six months because if the police get no new information from the prisoners, the evidence isn't enough to support conviction on a greater charge.

What would you do? If you can trust your friend, the best policy is to stay silent. That way, you do the least amount of time. If you cannot trust him, however, it's best to give him up before he does the same to you—that way, at least you stay out of prison. Danny and his brothers had clearly been raised in a world where this game was not a hypothetical problem. In fact, from their perspective, they may have seen their interviews with Bruce in this way. They were not going to give up anything, unless they thought he had something on them.

The prisoner's dilemma can help us understand altruism because of a mathematical outcome that becomes clear when the game is played repeatedly with multiple players who are allowed to react to the previous actions of their "friends." This can be used to create a computer simulation of a world where creatures have "reputations" and can reward or punish others' cooperativeness or betrayal in the next round. In this world, they reproduce or die based on which techniques work best. It turns out that if "defectors" (those who betray their friends) can be detected and punished, altruism can be part of a good

strategy for long-term survival. In populations with varying propor-
tions of always-selfish, always-cooperative, and "tit-for-tat" players,
tit-for-tat players can compete well. In many circumstances, these
"reciprocal altruists" don't die out or get swamped by selfish players.

Responding to kindness with kindness and betrayal with betrayal
can indeed be a winning strategy. In fact, even more successful in
some simulations are creatures that can get out of a Hatfield and Mc-
Coy revenge cycle by "forgiving" once in a while and returning to repeated cooperation if the other side responds in the same way. Because
in these simulations the players don't "think" or "plan" for the future,
they can represent animals during evolution who couldn't consciously
understand the potential benefits of "being nice."

Interestingly, new research finds that rewards may be even more
effective than punishment—at least for getting humans to cooperate
effectively. A study recently published in *Science* gave people a chance
to cooperate in what's called a "public goods" game. This is another
mathematical model that examines which strategies are most likely to
promote survival. In this version, everyone benefits most if group
members maximize their contributions to a pool of money or an
activity—but "free riders" can still benefit from the good done by everyone else even if they don't contribute at all. A real-life example
would be global climate change. We will all lose if carbon dioxide
emissions make the earth uninhabitable—but it's expensive or inconvenient to cut back or use alternative energy sources. However, if most
people manage to change, "cheaters" can benefit from their sacrifices
without making any of their own. How do we get them not to cheat?

As the research mentioned earlier suggests, one way is to punish
them—fines and prison sentences will do nicely. But enforcement can
be expensive—and widespread cheating can make people cynical and
untrusting, decreasing cooperation overall. The new data suggest that
rewarding cooperation is actually more effective. In the study, people

were given the opportunity to contribute money to a pool. The more people gave, the more money everyone would ultimately get. People also had the option of punishing stingy people—or rewarding those who were especially generous. When the game was played for several rounds, it turned out that rewarding cooperation yielded greater profits for everyone by increasing giving—while punishment did not increase giving and, therefore, reduced everyone's gains.[1] This suggests that humans may be able to develop more sophisticated methods of increasing cooperation rather than relying solely on punishment.

In nature, it turns out that social animals are very good at detecting whether others are "cheating" and taking more than they give. They seem to respond in exactly the tit-for-tat fashion that game theory suggests would be favored by evolution if animals had "cheater detection" and could punish offenders. A good, though gross, example is found in vampire bat colonies. When they feed, not every bat succeeds at bleeding prey. To help those who don't "score," the successful bats will actually regurgitate some of their own meal to feed their friends. But they keep track: and if one animal fails to help another when his friend is in need, the next time he's the one with the big meal, he won't share it with his selfish buddy.

Animals may have first begun to understand one another's intentions in predator/prey scenarios. A predator that can predict what her prey will do next will go hungry far less frequently than one that cannot; likewise, a prey animal that knows the next move of a predator will be better at escaping. But the ability to understand the intentions of others of the same species in a social scenario probably evolved out of the need to know who is doing their fair share and who isn't, who owes us a favor and whom we need to pay back.

These notions certainly have deep roots in our emotions, thought, and language. For example, even when you don't want to do so, it's hard not to "keep score" in friendships and other relationships. Just think about a recent conflict you've had or seen between friends. Chances are, somewhere in the dispute is an issue related to a failure of reciprocity. Typically, people become upset when they think that they "give more than they get." Alternatively, a friend can be overwhelmed by too much generosity and a resulting sense of indebtedness. Gift-giving occasions often unmask these kinds of resentments. And anyone with more than one child has frequently heard the phrase "That's not fair!" in relation to perceived inequitable distribution of goods and attention among siblings. It feels innate—parents certainly don't teach their children to monitor how much food and love their siblings get and to complain if they feel slighted! But siblings certainly do complain—often, even before they can fully articulate the issue.

The necessity of punishing cheaters means that the stable expression of altruism in a population depends at least in part on the ability to accurately detect "free riders." *Altruism can survive in a population only if those who don't do their part aren't able to get away with it for long.* Since the world is not static, strategies evolve to better detect and punish cheaters; meanwhile, cheaters evolve ways around them. Just as faster mice lead to faster cats, better safes ultimately yield smarter safecrackers. This "arms race" quality can help us understand why there are no societies that are completely altruistic—and none that is completely selfish, either.

In the quest for survival, however, what's best for the group is not always what's best for an individual, and so the conflict between these needs makes understanding the evolution of altruism quite complicated. Some traits may persist because they help individuals do better as selfish individuals within their groups; others may persist because altruistic traits help whole groups do better in competition between groups.

Certainly, the success of a social group often depends on the ability of its members to work together, so the balance between the use of selfish and altruistic strategies by its members is critical. A group made up entirely of altruists could thrive—and could even use its ability to work together to triumph over other not-so-coordinated groups. But a group of altruists with some freeloaders who take but do not give back to the group would not be as successful.

Here's where another of those evolutionary paradoxes comes in: a further driver of the development of human altruism and cooperation was probably the fact that the people who were best able to work together were most likely to survive because this allowed them to conquer others who were not as cooperative among themselves. In other words, our best qualities—self-sacrifice, kindness, cooperation—may have been the product of genetic selection because these traits helped our ancestors succeed at our lowest and cruelest endeavor, war. Indeed, research has shown that group bonds are strengthened by conflict with outsiders. A common enemy can certainly unite people.

And so, our empathy toward "people like us" probably developed at least in part because it allowed us to conquer "people like them." More hearteningly, another driver of the evolution of kindness was probably the need to cooperate in child rearing (see Chapter 8). Of course, a few selfish members can "free ride" in altruistic groups and take advantage of them to spread their less cooperative tendencies. But once there are too many free riders, the ability to work together is lost and everyone fails.

Danny's family was fascinating in this sense: they were literally a group of free riders and cheaters who relied on the kindness of altruists to run their scams. If other people weren't open and trusting, their "business model" would fail because checking their credentials or talking to former "customers" would reveal the rip-offs. Their own ability to work together was constantly being tested as well, however. Although they specialized in duping people they didn't know, if they saw an opportunity to get over on each other, they often took it. In one criminal

Traveler group, for example, a woman devised a scam to rip off Disney World by claiming she had been raped and beaten at one of its hotels on Halloween in 1992. In reality she had had consensual sex with a man—and her brother had later beaten her and then bound her to the hotel bed with duct tape. She sued for $3 million and was in line to receive a several hundred thousand dollar settlement when her sister dropped a dime on her and she wound up in prison instead.[2] Cooperation within these groups is clearly far from perfect.

Especially for children, this world of shifting alliances and inconsistent trust is terrifyingly unstable. Imagine being unable to trust your own parents or having to constantly try to figure out whether what you were being told by those around you was true or false. Obviously, there are many "normal" lies parents tell children in an attempt to protect them—but this goes far beyond that. Pretty soon, children like Danny become hyperalert to other people's motivations. Because the empathetic response of parent to child is the template for the child's development of empathy, the ability to care becomes stunted. These children have excellent theory-of-mind skills in that they can predict what others might think and do, but copying the deceitful adults around them, they use these skills to take advantage, not to help.

Unable to trust, such children become untrustworthy. As with all human traits, of course, genetic factors are involved as well. Indeed, over centuries, a group that engaged in this kind of antisocial behavior could even "select for" genes that reduce empathy, trust, and cooperation in the same way that a specific environment "selects" animals with genes most suited to it. For example, children born with highly empathetic qualities might tend to leave such groups when they become adults. Alternatively, those who join criminal families rather than being born into them are likely not to be especially empathetic, to put it kindly. Over time, these two trends could "breed for" the precursor traits for antisocial behavior, producing increasingly successful con artists.

There certainly are known genes that influence some of the characteristics Bruce saw in Danny and his brothers that could have been passed down along with the family's lifestyle. Danny's low heart rate and its failure to change the way most people's do when telling a lie are windows into the workings of his stress system. Genes and environment interact to determine its settings. For instance, one person might be born with a stress response system that is naturally overreactive, while another's system tends toward sluggishness. Raising the reactive child in a calm, loving home might produce a boy who worries some but is mostly happy and functional. If he had cruel and chaotic parenting, however, that same boy might develop severe depression and anxiety disorders. The child with a sluggish system might do fine in a calm home, too—he'd probably seek excitement, but it would be in socially acceptable venues like sports or the stock market. Raise that child in a chaotic situation, however, and that temperament could be dangerous.

Studies of children who lie excessively and show other disruptive and antisocial behavior frequently find that they have lower resting levels of the stress hormone cortisol. They also have lower skin conductance responses to stress and lower resting heart rates.[3] Although Bruce did not measure Danny's cortisol or conductance, it would not surprise us at all if his findings on these measures were similar.

Such children also tend to be much less fearful than others. One can think of several ways that this could reduce their ability to empathize. Most obviously, if frightening experiences aren't that distressing for them, they would probably believe that fear isn't too bad for other people, either—and therefore they would be less likely to avoid scaring or hurting people. Similarly, physical punishment would deter them less: if you don't fear being punished, you won't worry too much about violating rules. This feeds on itself in an environment where children do not get the early nurturing they need.

If a child doesn't get the affection needed to make the connection between people and pleasure, the pain of social punishments like a "time-out" or "going to your room" is also diminished. If you don't enjoy your family's company and don't care about disappointing them, you won't find being kept away from them unpleasant and won't be ashamed if you behave poorly. You can see how this would make ordinary discipline difficult. Most children raised with empathy want to please their parents. A child who doesn't will become increasingly difficult to manage. In the next chapter, we'll see more about how this actually plays out—both genetically and in the brain.

Another reason that an underresponsive stress system may be linked to lowered empathy is that this can also be connected with an overall lack of sensitivity to both positive and negative experience. If someone doesn't feel much of anything, he or she may seek out more intense experiences. This often leads children with a high tolerance for fear to take risks and seek stimulation, with less concern for the effects this might have on others. Again, if such a child is raised in a positive environment, these characteristics may not be problematic: explorers, mountaineers, scientists, doctors, and entrepreneurs all benefit from having genes that make them seek extra stimulation. But if children are raised without trust and nurturance, such genes can wreak havoc.

A whole body of literature on different types of antisocial behavior supports these findings. One study compared the sons of criminals who became lawbreakers themselves to those who did not: the sons who avoided crime had more responsive stress systems, as measured by higher heart rate and skin conductance.[4] Another study looked at girls with low cortisol. They were more likely to be impulsive, aggressive, and lacking in empathy than girls in the higher cortisol range.[5] Environments like the one in which Danny grew up can severely exacerbate these genetic tendencies. If a child faces frequent stressful situations and cannot trust that his parents will consistently help him negotiate them, he won't learn how to effectively modulate his stress system. In some cases, it's possible that chronic, uncontrollable stress could "burn out" certain aspects of this system: producing a paradoxically low resting cortisol level that isn't responsive to later stress. Ultimately, you wind up with a child who can beat a lie detector—and only relates to other people as objects.

THESE SPIRALING AND self-reinforcing genetic and environmental influences suggest ways that empathy and altruism can be affected not just in individual lives, but also in cultures and history. Children with the genetic potential to become empathetic and humane can be turned away from these positive qualities through developmental experiences—as in Danny's case, the chronic threat of violence in the home and the internalization of antisocial behaviors and values of a family. A child's brain will organize in a way that *appears* to be most adaptive to the world around him. And by age four, most of the brain's structural growth has taken place. If the young child's world is brutal, those adaptations will predispose him to behave that way as well.

A chronically activated stress system is exactly what you need to survive in an environment where life is nasty, brutish, and short: being hypervigilant helps you detect, evade, and cope with threat. Not trusting others is appropriate when they are likely to exploit you. Having a hair-trigger temper tends to instill fear in those around you, minimizing their willingness to challenge you. Although harsh child-rearing techniques are cruel, if you are going to live in a cruel world, they can be adaptive.

Consequently, children raised in an overcontrolling, misogynist home permeated with domestic violence will be likely to grow up to reflect that family's values and its style of violent problem solving.

Boys will be more likely to be abusive, brutish partners and parents. Girls will probably feel diminished and incompetent. They will be vulnerable to exploitation and domineering by someone "just like Dad." This pattern can be repeated, generation after generation. Despite the genetic potential to be thoughtful, compassionate, creative, and fully empathic, the dysfunction of this family system can continue. And worse: if the family is in a culture that promotes similarly violent and abusive practices, all of the other groups of that society, including other families, will take on those features and values. They, too, will solve problems with physical violence and see women and children as property. The chance to break out of that transgenerational cycle will be minimal, and the "transmission" of these destructive values and behaviors will be nearly 100 percent. However, if an abusive family lives in a society where women are valued, where domestic violence is discouraged, even criminal, its chances are much better. Social isolation, privacy, and intimidation can still allow these abusive practices to continue within a family. But the familial "transmission" to the next generation will be cut at least in half.

Although it might seem unlikely that culture could have such a direct effect on physiology and the genes that are passed down related to it, some intriguing research supports this idea. University of Michigan psychologists Dov Cohen and Richard Nisbett studied the reactions of college men to being insulted. They found that their reactions were not particularly affected by their size and strength or by their sense of emotional security. What mattered was where they came from.

When researchers bumped into people born and raised in the southern United States and called them "asshole," their levels of the stress hormone cortisol and of testosterone were much more likely to rise than when the same thing was done to their Yankee counterparts. The southerners were also much more likely to support the use of violence when presented with a scenario in which another man hit on their fiancée. And in the American South, there is a long history and tradition of defending one's honor through violence—as well as an ongoing elevation in violent crime associated with insults and "disrespect." Interestingly, other violent crimes—like murders committed during armed robberies where there is no element of disrespect—are not affected.[6] Culture and biology are deeply intertwined and interactive.

And so, the social world reproduces itself, in either positive or negative ways. Distress can make parents already inclined not to be empathetic even less so. Not only will frightening circumstances promote inhumane behavior, they will also tend to minimize intellectual capacity, as the fear that results from being under constant threat remodels the brain. Fear actually shuts down the higher regions of the brain, the areas of the prefrontal cortex involved in planning, creative thinking, and considering long-term consequences. Individuals who are always threatened cannot reason to their maximal capacity. When needing to fight or flee, too much reflection is not a good thing, so shutting down the contemplative part of the brain makes good sense when experiencing fear.

Again, this is adaptive only in the short term; in the long run, it means that when distress or threat pervades a society, its people's abilities to progress in areas that require such skills will be diminished. Trade, science, technology, art—and many other areas that require creativity and cooperation—will lag behind. It is easy to imagine how these neurodevelopmental principles could have been at play historically: "Dark Ages" could literally have resulted from violent, oppressive governance and kept societies that would otherwise flourish in a self-perpetuating state of violence and terror. This would have been fueled not only by traumatized, frightened people being in a more reactive, less thoughtful state, but also by cycles of vengeance resulting from previous attacks. As we will see in later chapters, raising children in a way that fully expresses empathy may be the key to cultural productivity, creativity, and security.

As a result, it is not surprising that most of our history is brutal and dark. Because, just as children grow up as a reflection of the developmental environment provided by their parents and immediate family, so does the inventiveness, creativity, and productivity of a people reflect the developmental environment of their society. There are, of course, many other factors that have an influence as well, but we think this developmental perspective offers critical insight.

One of the best examples of these influences can be seen in the different child-rearing beliefs of ancient Athens and Sparta. These two city-states existed concurrently in time—so the genetics of each of the Greek communities were arguably similar. Yet their values were very different and their child-rearing practices—at least that we know of for the sons of the upper class—reflected this variance. What endures of each of these ancient societies reflects their priorities.

The main legacy of Sparta rests in a word: *Spartan*. This word exists in English because of Sparta's notoriety for imposing harsh, even brutal, conditions on its people in order to be constantly ready for war. Though they did not fight battles themselves, even women in Spartan society were prized primarily for physical fitness. This was seen as increasing the chance that they would bear stronger sons.

Most of Sparta's populace actually consisted of slaves. Not surprisingly, they were shown little mercy. But even the sons of the ruling class weren't indulged. From birth, they were expected to show strength and cunning. Babies who showed any sign of weakness or birth defects were thrown into a chasm in the mountains at the direction of public officials. Those allowed to live were not coddled: baths were given in freezing water, and from the start, discipline involved shame, humiliation, and physical pain. Bullying was encouraged to stamp out weakness. At age seven, boys were taken from their mothers and put into training camps where they were not given enough food to satisfy their hunger. They were encouraged to steal to make up for this—but beaten, sometimes to death, if they were caught. Education was focused almost exclusively on discipline and military strength.

These extremes of deprivation, aggressive physical and weapons training, and intense competition helped create a very efficient ruling military class. Though they had no clue about the mechanism by which this worked, the Spartans were creating children whose stress response systems were perfect for short lives filled with intense conflict. These were brains designed to react aggressively, not think.

By contrast, in Athens, the sons of the ruling class were "spoiled." Their mothers and baby nurses nurtured them closely until about age six when formal education started. Even then, they still lived at home. Their education was a combination of music, exercise, military skills, philosophy, poetry, and basically, a more "rounded" exposure to the range of academics, arts, and physical activity then available. Conversation, debate, oratory, and the writing of plays and poetry was encouraged.

Athenian culture provided the foundation of Western civilization. The idea of democracy, the core debates of philosophy, the fundamentals of art and drama, and the beginnings of science are all traceable to Athens. Sculpture, poetry, architecture, and a host of other remarkable advances in culture were inspired, modified, or actually created by Athenians and their closest allies. By contrast, Sparta lives on primarily in that word for harsh conditions and its reputation for military conquest.

Though there are clearly many other rich determinants of the patterns of history, the way our brains are shaped by our early environments can have an echoing impact. The relational world in which we grow up influences our ability to connect to one another, the way we treat one another, and our response to power differentials. The impact of traumatic stress on the ability to think and plan well affects not just individuals, but societies.

Richard Hellie, a professor of history at the University of Chicago who died in April of 2009, used this neurobiological perspective to examine a period in Russian history, from the fifteenth through seventeenth centuries. This time was characterized by violence, oppression, and an overwhelming climate of fear. For hundreds of years, the creative output of this culture was virtually nil; little poetry, literature, sculpture, or other enduring products of human creativity and productivity are found. Extreme brutality was pervasive: torture was used to punish even nonviolent crimes, and religious dissidents were burned alive. Hellie argued that the cultural stagnation of this period reflected this widespread fear and trauma, which replicated itself as each generation continued to traumatize the next and build brains for reaction, not thought.[7] Other "Dark Ages" could probably be analyzed similarly.

In fact, researchers have long wondered why the earliest periods of human history itself are so stagnant. For thousands of years after modern humans evolved, it seems like nothing much at all changes. The same tools are used, the same artwork is seen; there's virtually no technological progress, at least that is visible to us in artifacts. It is possible that circumstances were simply too stressful and too dangerous during this time to allow our full capacities to blossom. No one really knows what circumstances changed to spur the start of agriculture and the growth of civilizations.

six | NO MERCY

SEVENTEEN-YEAR-OLD RYAN HAD EVERYTHING A teenage boy could want—and then some. A brand-new Lexus SUV. A huge house, wealthy parents, popularity. And now, his admission letter to an Ivy League school. It was time to celebrate. Ryan rounded up his friends, obtained booze, and planned a serious party. At first, it seemed like any other such event—excessive, loud, potentially dangerous for mixing drinking and driving and raging teen hormones, but not unusual in adolescent America.

Ryan, however, had brought in some extra "entertainment." On a whim, he had invited Amy, a fifteen-year-old developmentally disabled girl from the neighborhood. Some of his friends knew her, but she certainly wasn't one of the "in crowd" among whom Ryan was a constant presence. Amy's face lit up when he showed up in his car to invite her to party at his parents' pool. Maybe she'd finally have some cool friends! After that day, however, it would be a long time before Amy would smile, trust anyone, or even sleep through the night.

"We did her a favor," Ryan told Bruce later. By then, he was in an attorney's office, out on bail. Although he looked like a dream prom date, perfectly groomed, impeccably dressed, strongly built with bright blue eyes, there he was trying to explain why he'd raped Amy and made her "put on a show" for his friends. Bruce, as a potential expert witness in Ryan's defense, was examining him.

"I don't know what the problem is, really," Ryan added in a polite, well-modulated voice. "She never would have gotten laid by anyone as good as us."

RYAN DIDN'T SPEND his early life in a neglectful orphanage like Eugenia. He wasn't raised by a family of con artists like Danny. He wasn't beaten or starved or witness to domestic violence or wartime trauma. He grew up in a stable two-parent home. The teen didn't have any developmental disabilities or diagnosed mental illnesses. His IQ, in fact, was well above average. He certainly didn't lack for any material goods or suffer from living in poverty.

And yet, he was as cold as—perhaps even colder than—any sociopath Bruce had ever interviewed, including some killers. What could cause someone to be so lacking in empathy—so lacking in any sense of decency or shame—as to rape and publicly humiliate an intellectually disabled girl and make such a remark? What had gone so wrong? Was Ryan simply the proverbial "bad seed"?

He certainly had sociopathic features: that much was clear to Bruce simply from hearing that comment about his victim and knowing the details of the crime. Ryan had laughed while his victim cried; he hadn't paid heed to her signs of pain, fear, and embarrassment. He'd been the one who brought Amy to the party, who assaulted her himself, and who egged on the others in her public humiliation. He was clearly the ringleader. He'd thought Amy's pain and confusion was funny, nothing to worry about. Until the cops came to arrest him, he'd seen the whole thing as one big joke. He'd even boasted about it.

Sociopathy (also called "psychopathy") is not a psychiatric diagnosis, but it is defined by a complete lack of empathy and conscience.

Although all sociopaths can be diagnosed with antisocial personality disorder, they are at the extreme end of the spectrum of people with antisocial traits. All serial killers are sociopaths, but most murderers are not; and there are many sociopaths who do not engage in physical violence. Instead, like Danny's family, they con and hustle and manipulate without regard for the harm done to their victims. Most homicides are committed on impulse or in the "heat of passion"; the killer acts without thinking and later feels regret. But sociopaths kill or commit other crimes "in cold blood"—the way Ryan did. The defining quality is a complete absence of concern for anyone other than oneself. People really are objects to sociopaths.

Unlike the problems with empathy seen in autism, the deficits seen in sociopathy do not involve the perspective-taking or theory-of-mind capacity. As we saw in Danny's case, people with sociopathic traits may, in fact, have superior abilities in terms of mentally placing themselves in the minds of other people and predicting their behavior. This makes them extremely dangerous. Instead, they lack emotional connection, the caring that is at the core of what most people think of as empathy or compassion. From Ryan's story, we can learn more about the worst-case scenario that can play out when the development of empathy does not occur at all—and the consequences this can have for sociopaths and the rest of us.

Ryan's parents were, to all external appearances, upstanding members of the community. His sleek, blond mother, Amanda, had lots of nervous energy but no apparent mental health problems. She spent her days working at a worthy and important charity—and was active in fund-raising efforts to support that organization and other socially responsible causes. She was the second wife of Michael, a tall, brown-haired, and solidly built investment banker. Amanda and Michael appeared to have a good marriage. Despite the obvious stress of the situation they faced with their son, there were few signs of significant conflict. Like his wife, Michael was a philanthropist, and together they were frequently seen in the newspapers' society pages.

When Bruce interviewed the couple, both were extremely concerned about Ryan. They were horrified by what had happened at the party. Their main goal, however, was to protect him from further consequences. They were completely united on that front. To Ryan's parents, the whole situation was extremely distasteful, and they wanted desperately for it to go away. They would pay whatever it took to make that happen, they made that clear. No expense would be spared. They wanted Bruce to diagnose their son with depression or something—anything—that would show that it wasn't his fault so that he could go to college as planned, not prison.

So far, Bruce hadn't found anything in his conversation with Ryan that would suggest that the boy was depressed. He didn't see any diagnosable mental illness, only a probable history of lesser antisocial behaviors. The crime appeared to be part of a lifelong history of selfish, insensitive behavior, which had occasionally gotten him into trouble at school, but never for anything of this magnitude. Where and when did it all start? Bruce began to talk with Amanda about Ryan's childhood, searching for clues as to what had led up to this criminal behavior.

As he began the interview, Bruce learned that Ryan was Amanda's first child. The boy had two older half brothers. Mostly he didn't see them because Michael's relationship with his first wife was strained. Amanda herself had been an only child. She wasn't estranged from her family, but she didn't see her parents very often and had minimal contact with more distant relatives. Raised in an affluent family, she'd never had a teenage babysitting job or other work that would bring her in contact with small children. She hadn't been especially interested in babies as a young girl.

But when she got engaged at twenty-eight, she felt her biological clock was ticking. She became eager for a child of her own. Amanda and Michael had a fairy-tale wedding—and they immediately began trying to start a family. Ryan would be the first baby she ever really held or spent time with. Michael, too, had little knowledge or experience of young children. That ignorance would ultimately prove tragic both for Ryan—and for his victim, Amy.

Late in her pregnancy, Amanda began interviewing nannies to help her raise their eagerly expected baby boy. She wasn't going to leave any son of hers in some impersonal daycare center! Her childhood nanny had been trained as a nurse; she wanted someone efficient, who knew what she was doing. Her baby's nanny would live with them and be virtually always available. Amanda genuinely did want the best for her child—and she had the means to pay for it. She spoke with each candidate at length, in person. She discussed their qualifications and her impressions with Michael. As for herself, she intended to keep up her social life and her charity work. She felt that what children needed most of all from their parents was "quality time," she told Bruce.

Bruce inquired about the nanny's impressions of Ryan as a young child.

"Which one?" Amanda asked.

Ah, here we go, Bruce thought. This is interesting. It turned out that when Ryan was just eight weeks old, Amanda had noticed that he was far more affectionate with and responsive to his nanny than he was with her. His eyes brightened with joy when he saw the young woman; she was the one he smiled at. But his reaction to Amanda was much less enthusiastic. He almost seemed frightened. That didn't seem right, she said. She was his "real" mom, after all. And so, she fired the woman for being "overinvolved." Once more, she carefully scrutinized the new candidates, consulted with her husband, chose the woman who seemed best, and went back to her busy schedule.

But sadly, the same thing soon happened again. At one point, in fact, Ryan wouldn't even let his own mother hold him and screamed and struggled until his nanny returned. He was about nine months old. This made no sense to Amanda. She couldn't understand what was wrong and why her own baby didn't seem to like her. She fired the second nanny. Clearly, these women were turning him against her somehow. She couldn't figure it out. Maybe they were jealous of her privileged lifestyle?

Bruce questioned her more closely, trying to get a better sense of what Ryan's early life had been like, day to day. However, he soon found that Amanda couldn't tell him much. She didn't know many specifics or details about his schedule or activities. Over the course of the conversation, Bruce began to discover a terrible truth. First, he learned that Amanda and Michael had a very unusual notion of "quality time." For her, that meant an hour, once a day, reading and a little snuggle and kiss before breakfast and at bedtime—that's it. Day and night, baby Ryan was cared for by the nanny. Michael sometimes spent time after work playing with the boy, but his job often kept him out of town or very late at the office. Those short intervals were the only time Ryan really spent with his parents.

They simply weren't aware of just how little contact that really was in a baby's world. Amanda and Michael genuinely believed they could be close to Ryan that way. They didn't realize that frequently repeated routines—almost endless repetitions and specific daily rhythms—are the heart of parenting. Bruce could clearly see the origins of Ryan's problems with empathy as he continued the interview. Cringing inside, he learned that by the time Ryan had turned three, the family had gone through eighteen nannies.

By then, Amanda told him, Ryan had seemed much calmer. He didn't scream anymore when his nanny wasn't around. He no longer protested when Mom held him. He didn't cry much at all. This was something that Bruce had seen frequently in other children with disrupted attachments: after a while, they just stopped crying and seemed to give up on getting their emotional needs met. Ryan seemed to enjoy the books and toys that his parents bought for him. Although Amanda and Michael admitted that there had been "occasional" complaints from school and from neighbors about "minor" bullying and vandalism, they minimized these and described him as a good boy. He was a great kid, both parents insisted. There was never any hint of this sort of behavior, nothing. And his academic and athletic achievements spoke for themselves.

Bruce now knew, at least in large part, what had gone wrong. Eugenia had experienced caregiving in shifts at the orphanage—and this had clearly been quite disruptive to her development. After being placed by her mother, she didn't really get the chance to connect with anyone else for over two years until she was adopted. At the orphanage, her caregivers rotated, but mostly, they didn't completely disappear. Once she was adopted, she had two parents and a little brother to connect with.

Ryan's early experience was different, but, sadly, in many ways it was much worse. Unlike Eugenia, baby Ryan got to spend plenty of time with each caregiver, almost all day and night, every day. Unlike her, he got focused, intense individual attention. But then, just after he'd learned one nanny's scent and touch and smile—as soon as he had begun to bond—she'd disappear. He'd get familiar with another "mother." He'd start to get a sense of her rhythm and voice and style. And whoosh! Gone, once again. From his perspective, he'd been abandoned by eighteen "moms," all before he finished preschool.

He had learned that every time he got attached to someone, she would soon leave. The lesson was that there is no real safety or comfort in people. His protests and complaints, his crying and yearning to see his familiar nanny again, were ignored. They weren't even recognized. Before he even started kindergarten, the relational part of his brain had become stunted and it functioned abnormally. There's no way of knowing what genetic influences he brought to the situation, of course, but this repeated loss of attachment is enough to be devastating to child development.

Imagine the impact on speech and language if a baby's primary exposure to words was to hear eighteen different languages in a row, each only for a few months, never to be heard again—all before he was three. The relevant brain regions would be a disorganized mess. And his speech and language skills would be immature, disorganized, and, essentially, nonfunctional. The same thing happened to the relational parts of Ryan's brain. His capacity for empathy was underdeveloped and immature (he was selfish), it was disorganized (he got no pleasure or soothing from reciprocal social interactions), and ultimately, it was nonfunctional (he was incapable of being empathic).

As we've seen, empathetic capacity develops first in the early bond babies make with their primary caregiver. It doesn't matter to babies whether this caregiver is their biological mother or father or a nanny or a grandparent: what matters is consistency and nurturing. Oxytocin and dopamine and endogenous opioids do their things, a child bonds, and his or her stress response system grows. In circumstances like Ryan's, however, repeated partial bonding and loss prevents secure bonds from developing. Over time, it becomes too painful for a baby to even try to connect again. That part of him simply shuts down. Rather than trying to find pleasure and comfort in people, he turns to things. And Ryan had access to many, many things. Those shiny objects couldn't help, however, with the changes that his early life had wrought in his brain—and even in his genes.

IN LATE PRENATAL life and early infancy, babies' brains are searching for clues about the world that they are entering, trying to predict what settings will best adapt them to the life ahead. Again, this is a feature, not a bug—without being able to do this, many species would not have been able to survive the changing environments they faced. A baby is born with just one suite of genes. If they can produce only one pattern of responses, that pattern could turn out to be fatally maladaptive. But if genes are programmed to be "set" differently by different early environments, organisms can be much more flexible. A whole new field of science—called "epigenetics"—has been developed to study these interactions between genes and the environment. This research is showing us that these settings involve changes in the way our genes themselves work.

From Ryan's brain's perspective, he was born into a chaotic world of loss, despite all the material advantages he'd been given. Each time he thought he'd figured out how to connect and calm himself, the person who allowed him to do that comfortably went away. Each association he made with safety disappeared, never to be experienced again. Though it didn't look like it to Amanda and Michael, they'd created a terrifying, extremely stressful emotional environment for Ryan. And it wasn't one that would enhance his capacity for empathy.

So how can early experience be powerful enough to affect genes? The most intensive research on this has been done by Michael Meaney, director of the Program for the Study of Behaviour, Genes and Environment at McGill University, who has examined the details of this process in rats. He's studied two primary mothering patterns in these rodents, both of which help determine which fragments of DNA are used to guide development and behavior—and which are not.

To understand how this works, we need to look a bit more closely at how genes affect the brain and body. Genes are basically recipes for proteins. These proteins do much of the work of life: everything from serving as parts of cells to regulating digestion and sending messages from one brain cell to another. They make the brain and body function.

When a gene is activated to produce a protein, it's called "gene expression." But if a gene isn't expressed, it has no effect. What Meaney's work would show in detail is that a baby's early environment determines how and even whether certain important genes will be expressed throughout his life. For mammals like us, of course, the most critical part of the early environment is one's primary attachment figure—usually, the mother.

That's right: the way that a mother treats her baby early in life literally affects which DNA gets transcribed and, therefore, the physiological path the baby's brain and body will take. This is where nature meets nurture. Here, nurture determines which natural potentials—positive and negative—will be realized and which will stay hidden in the double helix.

When he began his career, Meaney was torn between becoming a biologist and studying psychology. He wound up at their intersection, studying how they affect each other. He'd seen research showing that early life events could permanently change rats' stress responses. In fact, that research had been done by one of Bruce's mentors, Seymour Levine. It had also had a huge influence on Bruce's thinking about these issues. Levine's work found that just a few moments of handling of baby rats by humans could change the rats' stress response systems for life. Meaney set out to discover: How could such small events have such a profound effect? What was the mechanism?

He knew that some rat mothers were far more affectionate with their babies than others. These attentive moms spent more time licking and grooming their pups and more time "arched-back nursing." This nursing position is likely the rat equivalent to a human mother gazing into her baby's eyes and paying attention while feeding him, as opposed to focusing mainly on other tasks. When rats nurse lying down or without their backs arched, they groom their babies less. More arched-back nursing means more nurturing. (Note to moms: we are not in any way suggesting that every moment of nursing needs to involve your full attention. Even the most nurturing mothers—rat or human—don't do this!)

And as we've seen, being groomed is critical to a baby rat's health. If a rat pup doesn't get any licking, it will die. Eugenia's story further illustrated how physical affection is necessary for growth hormone production. But Meaney discovered that the rat pups who received extra attention showed measurable changes in their brains, too. Not surprisingly, they grew faster and healthier. The females had more oxytocin activity in the brain's pleasure regions—like those highly social, maternal, and monogamous prairie voles. Rats of both sexes raised by more nurturing moms were less anxious when placed in an open field, a situation that is normally scary for them because there's nowhere to hide. These rodents were also smarter in most situations— better able to negotiate mazes, find hidden platforms in water, and recognize objects. And their stress systems were calmer.

Importantly, Meaney found that rats raised by the highly affectionate mothers had an enhanced GABA-A receptor system. In part, these brain cells act as a brake on the HPA axis (discussed in Chapter 2) of the stress response system, making it less likely to overreact. This is where benzodiazepine drugs like Valium and Xanax have much of their effect. They reduce anxiety by chilling out the HPA stress response, calming us down. Basically, the more nurturant moms gave their babies a greater ability to soothe themselves. Consequently, when they were put in stressful situations like being restrained and unable to move, their stress response was less likely to spring into overdrive.[1] Other changes affecting this system included decreased numbers of receptors for CRH, a hormone that turns the whole system on.[2] And so, not only did these highly nurtured rats have stronger "brakes" on their stress responses, they also had a "starter" that was less likely to turn them on in the first place.

Now you might think that this was really a genetic effect. Perhaps the babies of more affectionate rats had inherited their mothers' relaxed, nurturing genes. Maybe this was why it looked like extra licking made them calmer, when they were really just predisposed to be mellow in the first place. Meaney recognized this objection. He countered by "fostering" the babies of high-licking moms to low-licking moms—and vice versa.

Would he see the same contrasts? Certainly, it was possible that these differences would be muted or even eliminated due to genetic variations. But surprisingly, the outcomes were exactly the same. There was no measurable genetic effect. None. The babies of low-licking mothers given to high-licking mothers were just like the natural offspring of the high lickers: calm and collected. They showed all the same changes in their stress systems and GABA-A. It wasn't a genetic predisposition; it was mothering that was changing their brains. And the babies of high-licking rats given to low-licking rats behaved like their foster moms, too.

Then the findings became even more fascinating. As it happens, extra nurture doesn't just affect babies' behavior during childhood and adolescence. In females, it also changes their maternal style when they have their own offspring and, thus, affects the behavior of the next generation. When these baby rats became adults, their mothering behavior was not like that of their biological moms—it was like the nurturing style of foster moms! If a rat was raised by a high-licking mom, that's the kind of mom the baby rat became. And the same was true for their babies, too. The grandchildren of the original rat mothers took on the maternal style with which they'd been raised. In humans, this wouldn't be a big surprise: we tend to parent as we were parented. But how could rats—which don't share genes with their foster parents and grandparents and don't learn by imitation—pass a mothering style on to the next generation?

It turns out that mothering and other environmental factors can and do affect DNA. DNA can be seen as a recipe book. Each gene or recipe makes only one type of "cookie." But if you don't decide to bake, chocolate chip cookies don't simply arise by themselves from your cookbook, sadly. Meaney's research found that the mothering style of a rat affects which DNA gets transcribed—which recipes the cells decide to "bake" and how many "cookies" will be produced. These transcription changes are affected by the environment and passed down by influencing the environment of the next generation. And it is through these changes that more maternal rats produce calmer, more nurturing babies and grandbabies—and less maternal rats do likewise.

Consequently, it seems that at least part of the way we "remember" how we were parented early in life is expressed in our genes (it is important to note that this does not change the genes themselves; it only influences which ones turn on and off as we develop). When we become parents, then we will "instinctively" tend to do it the way our own mom and dad did. Of course, for humans, it's much more complicated than for rats—and we can consciously choose to do things very differently. Nonetheless, this research suggests one reason why it may take more effort to change your parenting behavior ("I'm turning into my mother!") than it is to alter other habits. New styles of parenting behavior may not feel "natural."

Importantly, the gene expression patterns seen in rats born to less nurturing moms aren't all bad. They are adaptive for highly stressful environments. Extreme stress, in fact, tends to make all mothers—human or rat—less affectionate toward their offspring. But this can serve a purpose. The rat research finds that under overwhelming stress, pups born to less nurturing moms are smarter, at least in terms of solving threat-related problems. Their memories for frightening situations are better. These qualities help them survive, at least in the short term. Their brains don't just show deficits—they have certain regions where performance is enhanced.[3] Essentially, less maternal moms are better preparing their babies for a stressful, frightening, and short life—whereas warmer moms are readying their children for a kinder, gentler planet. And there's growing evidence that these changes apply to humans, too.

By putting him through so many changes of caregivers, Ryan's parents had unwittingly created the equivalent of a "low-licking" high-stress environment. Unsurprisingly, this stressful situation made him less likely to be empathetic. Failing to experience normal bonding changes the oxytocin system. And this can have terrible consequences, particularly with regard to the ability to find comfort and pleasure in loving and being loved by others. Lack of nurture directly affects key brain regions that create the experience of joy, desire, and motivation. Meaney's studies found specific differences in these rats' "reward" systems related to their early environment.

The highly nurturant mothers got a bigger hit of dopamine—and had higher levels of it for a longer period of time—just before they groomed their pups. This dopamine response reflects an expectation of pleasure. It's seen intensely, for example, in addicts yearning for their next hit of drugs; the strength of an addict's craving is directly correlated with the dopamine spike.[4] Essentially, the more nurturing moms were more driven to nuzzle. They found grooming their babies more delightful and rewarding.[5] That made them do it more. Like addicts, they craved—in this case, they just loved loving their babies.

The most interesting findings were in the offspring, however. As with the changes in oxytocin, these differences resulted from the mothering the babies received, not from their genetic backgrounds. Rats raised by the most affectionate moms were more likely to find licking their own pups rewarding. They wanted to do it more, again. The others just weren't as keen. But not all types of pleasure and craving were equally affected. In fact, rats raised in isolation got "higher" when given cocaine[6]—and rats that received less licking after being away from their mothers were more likely to repeatedly take cocaine and alcohol than other rats.[7] Similar results have even begun to be seen in human imaging studies, with research finding changes in responses to other stimulant drugs like Ritalin related to an early childhood history of either highly attentive or inattentive parenting.[8] Your early environment—specifically, the amount of nurturing you receive—can determine how much pleasure you get from nurturing, from being nurtured, and from drugs. This is how nurture implements nature.

THE IMPLICATIONS HERE are profound: childhoods like Ryan's or Eugenia's can change the brain in predictable and sometimes devastating ways. The most troubling are the alterations in the connections among oxytocin, dopamine, and pleasure. These connections reflect fundamental human capacities: the ability to take joy in loving and being loved, and the ability to find happiness in pleasing others and being pleased by them and by your connection to them. The changes caused by disturbances in early attachment physiologically make social interactions less attractive, less pleasurable, less comforting. They affect not only how rewarding we will find relationships with our parents and friends—but also, how we will tend to parent our own children and how much joy we will easily take in them. Essentially, they help determine how good it feels to love and be loved—and that, of course, can affect virtually every aspect of life.

If bonding with others doesn't feel particularly compelling, there's less reason to be kind or to follow social rules. If you don't care much about relationships, why would you worry about whether you hurt someone's feelings? In fact, if you don't care about connecting with people, your only motivation for interacting with them will be instrumental: you'll do what they want only insofar as it helps you get what you want. People become interchangeable objects, to be manipulated not considered as equals.

As mentioned earlier, these changes can also make social punishments like a "time-out" less effective. Being less loved—or having the repeated early experience of loss like Ryan—can also make loving itself harder and less satisfying. Like an addict with a tolerance, it takes a higher "dose" to get the same effect.

Consequently, neglected children or those with other attachment disruptions are much harder to soothe or to teach: it takes a great deal more attention to calm them down and make them feel better. Some become incredibly needy—like the child in the classroom who clowns or cannot work unless he receives constant praise and guidance for staying on task. A child like this isn't being selfish; it is simply the case that each little dose of affection has a smaller, less lasting effect on his brain. As such children grow up, these tendencies can obviously lead to problems in relationships as well—sometimes producing a clinginess and desperation that often drives partners either crazy or away. Fortunately, as we will see in the next chapter, some children learn to manage the results of such early experience in highly creative, loving, and productive ways.

Other victims of this kind of childhood—like Ryan—seek pleasure elsewhere. Not finding relationships rewarding, they seek thrills in physical pleasures and in wielding power over others. The buzz they get from food or sex or drugs isn't diminished by these epigenetic settings—it actually seems to be enhanced. Only their comfort in relationships suffers. This has major public health implications because it increases every other hunger, dramatically raising their risk for all addictions and compulsive behaviors. Whether it's cocaine, heroin, methamphetamine, or alcohol; sweet, salty, or fatty foods; gambling, video gaming, or promiscuous sex—anything that can be an escape attracts them. Though Ryan wasn't an addict or alcoholic when he committed his crime, he certainly was at greater risk of becoming one than his unaffected peers were. His callousness reflected a kind of relational blindness that was shaped by loss of connection.

In this way, chronic, uncontrollable stress early in life can actually change gene expression—and these changes can be passed down from one generation to the next through alterations in parenting behavior. Being prepared for a stressful world increases aggression, while being prepared for a calmer world increases love. Again, this has huge implications. It means that the traumas of war, genocide, famine, and natural disaster can linger and be passed on, affecting history through epigenetic influences. It is a mechanism by which the sins of the father can pass to the grandchildren and beyond.

For Ryan, being denied the repetitions and consistent closeness needed for bonding diminished not only the capacity to empathize with others' pain, but also the capacity to share their joy and to find relief in their company. Ironically, though he was surrounded by wealth, Ryan's brain had inadvertently been adapted for a world in which, essentially, it was every man for himself. He was financially rich—but relationally, far below poverty.

Please note that we are not saying that this excuses his behavior. Many children react to neglectful or abusive situations with greater desire for and interest in empathy. Early attachment problems can also be mitigated by strong, later-life relationships. Children themselves make many choices during development that help determine a life's path. Some of these are conscious and deliberate, some less so— all of them add up to a self that includes moral agency. Complexities certainly arise in the attribution of ultimate responsibility because many small, unconsidered choices can wear in a groove that makes whatever direction a child starts in, positive or negative, very difficult to change. And external factors over which someone has little control—like parenting and the presence or absence of warm, kind people when hope is most needed—can, of course, make a huge difference.

That said, Ryan's coldness didn't come from nowhere. In fact, it was exacerbated by the world of riches that surrounded him and the culture of his community, which glorified money and status. His family had used its wealth and influence for years to pressure other families, the school system, and his coaches to minimize the consequences of his selfish, hurtful behaviors. They had actually facilitated the growth of his disturbed personality before his crime unmasked him. The way those factors combined to allow his sociopathy to remain undetected for so long is another important part of his story—one that has serious implications for the development and sustenance of empathy in America.

Ryan had been genetically blessed not just with privileged parents but also with above average intelligence. Intelligence is one quality that is usually linked with successful psychological development despite severe stress—as is clear in Eugenia's case and in the story of a remarkable woman we'll meet in the next chapter. Unfortunately, it did not help Ryan develop empathy. Instead, he used his intelligence to avoid consequences. His brain's cognitive regions had received the stimulation they needed virtually from the start—and regions that get what they need develop, while others wither. His mother read to him. He was exposed to lots of language and toys and educational activities. He went to an excellent preschool and kindergarten. By this time, his mother had stopped firing nannies so frequently and his relational world stabilized.

Because he was smart, Ryan easily recognized what was expected of him. He developed perspective-taking skills on time, possibly even early. The problem was that when he put himself in someone else's position, he didn't do so to understand or connect with that person. He did it to discover the best way to manipulate him or her. His own early emotional needs had been ignored. He didn't really recognize anyone else's. He couldn't take much pleasure in relational connection, so he didn't think others really did, either. From his point of view, for example, something like professing one's love for a girl was just a lie to get sex; that's the way he thought about it, so that's the way he assumed others would, too. All that lovey-dovey stuff wasn't real; it was just a way to hide your true motives.

Consequently, Ryan's interactions were entirely about winning or losing. If there was something he wanted to do and his parents or teachers forbade it, he would either find a way to do it without being caught or talk an adult into giving permission. If he did get caught, being lectured or punished didn't have much effect: it simply didn't bother him if he displeased people—or pleased them, for that matter. It was all a cynical game. Friendly ties were just alliances of convenience—no relationship was genuinely reciprocal. Seeing the rest of the world as selfish via perspective taking even allowed him to rationalize and further justify his behavior. It was a dog-eat-dog world, that's all.

Ryan also rapidly picked up on his family's high status and their sense of entitlement. His house was one of the biggest in the neighborhood; his family's cars were among the most expensive. He could have the latest clothes and music pretty much whenever he wanted them. He saw the power that his parents wielded and watched how it affected less privileged people like his nannies and the other household staff. He knew that he never wanted to be on the other end of that power dynamic.

Ryan recognized that his parents valued academic and athletic achievement—so that's what he gave them. Since those things weren't difficult for him and he mostly enjoyed them, it was no skin off his back. He also knew how to do just enough work to get the grades he needed; if he slipped up, he could always cheat. And he also learned that, conveniently, as long as he did well in those arenas, he wouldn't be watched too closely. In fact, he would get a pass for most misbehavior. The rest, he was smart enough to hide.

And so, when he bullied other children, it was brushed off as "boys will be boys." When he targeted the class scapegoats, it was their fault for being nerds and having laughable social skills, not his. Being among the richest in his class, being good at sports, and knowing how to work people made it easy for him to become popular; he knew how to make people see him as a winner. The dominant culture at his high school did not discourage such attitudes. Teachers and coaches themselves would sometimes even laugh along when Ryan and his friends put down the less popular teens. Little effort was made by school officials to promote tolerance and inclusion.

Bruce had seen this before when he'd been asked to help develop a community response to high-profile school shootings like Columbine. There was a pattern. Although schools that experienced shootings or crimes like Ryan's did honor academic achievement, athletic performance was given outsized attention, financial support, and social prominence. Athletes were treated as "special" and often their bullying, cheating, or rule-breaking behaviors were tolerated or ignored. Other activities like music, drama, and debate were given short shrift. Community service was something that looked nice to college admissions officers but wasn't universally encouraged. Status was created by having specific clothes, expensive jewelry, or cars. Outsiders in these schools were often the butt of jokes, teasing, and overt physical intimidation. The adults ignored the climate of fear this created for the scapegoats, who were on their own in dealing with the "Ryans" of the world. Not a climate to promote empathy.

A similarly horrifying case and the culture that produced it were eloquently described by Bernard Lefkowitz in his book *Our Guys: The Glen Ridge Rape and the Secret Life of the Perfect Suburb*. In this New

Jersey town in 1989, a clique of popular athletes raped a developmentally disabled girl with a broomstick and a baseball bat and forced her to perform oral sex. As in Ryan's case, some of the school's most popular athletes were involved in the crime. Similarly, there was widespread community sentiment that even if the crime had happened exactly as the victim claimed, the boys' futures were too valuable to be squandered. Because the victim was a disabled person, she wasn't seen as likely to have a "future." She wasn't "worth" destroying the golden lives of the town's "scholar athletes." As one parent put it at a graduation party attended by the suspects in the case, "It's such a tragedy. They're such beautiful boys and this will scar them forever."[9] Unmentioned was any notion of the boys' responsibility for the "tragedy"—or any concern for their victim, who was almost certainly more likely to have lasting scars than the perpetrators.

Bruce suspected that in Ryan's case, there were many previous incidents of serious cruelty or violence that had either been dismissed by authorities or never came to their attention because he was clever enough to either hide his behavior or intimidate any victims or witnesses into keeping quiet. This had certainly been true in Glen Ridge. There, one of the perpetrators had frequently masturbated publicly and exposed himself to girls during class, starting in middle school and continuing throughout high school. His sexually predatory behavior was common knowledge—even some teachers knew and did nothing to stop it. And the other boys—just like in Ryan's case—were at an age at which they were uniquely susceptible to "following the crowd," even when its leader instigated actions that they would find morally unacceptable on their own.

Many of the Glen Ridge perpetrators had also been involved in a drunken three-day rampage that rendered uninhabitable the house of an unpopular girl whose parents had gone away for the weekend. During the incident, partygoers seriously burned the girl's cat by putting

the unfortunate animal in a microwave oven. Most of the family's fur-
niture and possessions were shattered and trashed. When the girl
threatened to commit suicide by jumping out a window, the other
teens chanted "Jump! Jump!" Thousands of dollars' worth of damage
had been done—not to mention the emotional toll of such willful destruction and cruelty. The incident drove the family out of town. But
no one was punished; no one even apologized. These youths were
proud of their participation, not chastened. Indeed, several of the perpetrators listed the event as a highlight of their high school memories
in the yearbook. Similarly listed was an activity called "voyeuring" in
which the athletes watched one another's sexual behavior, often spring-
ing out of closets to catch the girl involved unawares and humiliate
her.[10]

Lefkowitz observed that only one of the boys charged in the case
had any sisters; all had spent most of their time on all-male teams and
had little experience of seeing girls as equals. Even in class at school,
the guys stuck together. Studies find, in fact, that gang rape and simi
lar crimes are more common in cultures where men and women are
frequently segregated and women's roles are viewed and enforced as
inferior.[11] Lack of exposure to particular types of people and extreme
differences in status are both factors that dramatically decrease empathy for those groups. Schools that don't work to highlight gender
equity, that don't focus on inclusion, and don't try to minimize all
forms of segregation risk decreasing empathic behaviors and indirectly encouraging bullying. Though sports teams can often incul-
cate positive values, if they are promoted and idolized uncritically,
they can instead enhance an "us-versus-them" view of the world.

Sex crimes like Ryan's are more commonly reported among
athletes—virtually always those who play team sports. But it is not the
outcasts or low-status men who tend to be involved: one study, which
looked at twenty-four known cases of gang rape on college campuses

between 1981 and 1991 found that members of elite fraternities or top sports teams tended to be the perpetrators.[12] These offenses don't come from "low self-esteem" or powerlessness; they come from an overwhelming and callous sense of entitlement, from rewarding performance while ignoring character. Although some of the perpetrators may be otherwise decent boys who get caught up in group pressures to conform, they follow young men like Ryan who have no moral compass. And they are vulnerable to doing so at least in part because as a society, we have let the stock of empathy decline, while emphasizing winning at all costs.

RYAN'S PARENTS HAD never been educated about the social needs of infants. Unfortunately, even today, virtually nothing is taught about this in high school or college, unless psychology or child development classes are deliberately chosen. And, as an only child, Amanda had no younger siblings to play with and learn about babies. She'd spent little time with extended family, friends, or neighbors with young children. She wasn't close with her own mother, who lived hundreds of miles away. Michael grew up in similar circumstances. The age segregation of our country is increasingly problematic—reducing empathy for and understanding of both the very young and the very old, just as sex segregation can reduce empathy between the genders. We can't empathize well with types of people we just don't know. If we don't know what's normal for a child's stage of development, we won't have appropriate expectations and may fail to meet his or her critical needs.

Clearly, such "child illiteracy" isn't limited to the poor: the wealthy and middle class can sometimes be even more ignorant about child development. Smaller nuclear families, busy overscheduled professional parents, and high mobility combine to create an environment in which the affluent may spend less time than the poor around babies and children. Although this is unprecedented in human evolutionary history, it is now not uncommon for women—like Amanda—to have their first real experiences with babies when they have their own. As we've seen, such ignorance can cause serious damage.

And despite recent media campaigns and reporting on the importance of early nurturance and bonding, widespread child illiteracy persists. In a 2007–2008 child custody case, for example, a judge decided that it would be OK to take a one-year-old child out of the only home she had ever known—even though by all accounts she was receiving appropriate, even exemplary parenting—because her foster parents were lesbians. As *New York Times* writer Pamela Paul described it in her article "The Battle Over a Baby," West Virginia Circuit Court Judge Paul M. Blake believed that, "Uprooting an 11-month-old baby, while not ideal, wouldn't be traumatic. Who among us remembers what happened when we were a year old?" In fact, all of us do—but these memories aren't conscious, simply foundational. They program our genes and our brains to face a particular type of world; they calibrate the degree to which we will take pleasure in relationships. They matter deeply—as Ryan's story tragically exemplifies.

By the time someone understood the implications of Ryan's disrupted attachments, it was too late. For once, he couldn't entirely avoid the consequences of his actions. He was convicted of crimes related to Amy's rape. He served prison time, studying business while incarcerated. After his release, he went to work for a firm owned by a family member.

Thankfully, few people will ever suffer the type of repeated early loss that he did—and fortunately, even most of those who do will not become as cold and hurtful as Ryan. Nonetheless, we ignore the emotional needs of young children at our peril. Mix child illiteracy with an individualistic culture that promotes competition instead of collaboration, add a mélange of electronic media that can be isolating and violent and can reduce time spent socializing, and you create a world where empathy is threatened. As we'll see in the next chapter, although empathy can sometimes overcome even the most harrowing childhoods, many elements are necessary in order for it to flourish. If we don't provide them, we risk creating a society where everyone is just a bit less connected to one another, where we all find nurturing one another just a bit less rewarding. A world where more people think and behave the way Ryan did—and more people parent like "low-licking" rats.

seven | RESILIENCE

W HEN TRINITY WALLACE-ELLIS was in elementary school, she remembers arguing with her teacher, Miss Olive. It was the mid-1980s. Nancy Reagan's "Just Say No" campaign was in full swing. Trinity was eight or nine, living in inner-city Los Angeles, attending one of the seventeen schools she'd pass through before she graduated high school. Drugs were a subject that had her full attention. The teacher said, "Parents who do drugs have children who do drugs." Trinity immediately and urgently raised her hand. When called on, she insisted that that wasn't true. "Just because parents do drugs doesn't mean their children will," she affirmed.

Miss Olive repeated that this was what the statistics showed, but she had lost Trinity. She hadn't picked up on the fact that this was an issue of great personal interest to the little girl. She didn't consider what that statement might mean to the child of an addict: the anti-drug program hadn't been designed for "those" kinds of children. The teacher—and the program—had failed to empathize with the students who most needed to be reached.

Miss Olive had no idea that Trinity's father was a heroin addict, a drug dealer, and a pimp. She didn't know that Trinity often helped her dad "fix," or that her mother was also heavily involved with drugs. She

wasn't aware that Trinity and her seven younger sisters were frequently home alone, unsupervised with little or nothing to eat. Nor did
she recognize that Trinity had sworn to herself that she'd never be like
her parents.

"I'm not a statistic," Trinity told her teacher—and, more important, herself. Later, she would hear the same kinds of statements
about abused children being fated to become abusers themselves. But
she didn't let herself buy that, either. The insensitivity of the world
around her wasn't going to get in her way.

TRINITY WAS RAISED in a highly stressful environment, frequently a
victim of multiple types of abuse and neglect. But this isn't a story of
the child of an addict who became an addict or a child in foster care
whose children were lost to foster care. Her story is nothing like Ryan's. It is instead a tale of resilience in the face of great challenges. It
illustrates the complexity of the development of compassion. Although
the capacity for empathy relies to a large degree on a child being the
recipient of empathy, in some circumstances, even small experiences
of kindness can go a long way. When the historical cycles of violence
mentioned in Chapter 5 take hold, the people who can remain empathetic in the face of them are a precious source of hope and revival.
Trinity Wallace-Ellis is one such person.

When her father, Mohammed, was clean, he was loving and
intelligent—but while high, which, unfortunately, was most of the time,
he was unpredictably violent, impatient, and threatening. Her mom,
Cassandra, wasn't primarily addicted to drugs. Instead, she was
hooked on abusive men, whose drug habits she would emulate when
they were together. It wasn't a healthy combination for children, to say
the least.

Trinity's father did take great pride in naming his children. Her name was chosen because she was born on the third day of the month at 3:13 in the morning in the presence of three people. Trinity's sisters' names include Righteousness and Eternity and other similarly spiritual concepts. Each also has a personalized and creative derivation. But unfortunately, Trinity's parents took far less care with their children's lives than they did with their names. Says Trinity wryly, "My parents are the smartest dumbasses I've ever met."

Mohammed had eight children with Cassandra and nine with other women. Meanwhile, Cassandra also had children with other fathers: she had ten on top of the eight who included Trinity. As the first of eight daughters the couple bore together, Trinity became, essentially, a second mother, a de facto parent to her younger sisters and to older siblings as well because of her intelligence, maternal nature, and reliability. "Even though I'm smack in the middle, I'm like the oldest sister even to my older brothers," she says.

Trinity's story shows the incredible resilience of the development of empathy—and how even on what seems like the rockiest and most unforgiving ground, this capacity can sometimes bring forth love in abundance. She's now thirty-one, a successful advocate for children in foster care, recently serving as director of Child Advocates of San Bernardino County, a California organization that provides trained volunteers to guide individual foster children through the courts and stand up for their needs and interests. She's married, with two children of her own—and has never been in jail or addicted to alcohol or other drugs. These days, she's also caring for her youngest brother, who is just one and a half, and she continues to help her other siblings.

"People tell me all the time, 'Trinity, you need to let them go,' and I say, 'I can't do that.' They need somebody, everyone needs somebody who knows—and I'm the one who knows what happened to them. They're not bad people. They've just made some bad decisions." All of her siblings except for Trinity have had run-ins with the law (though none has committed violent crimes). Two sisters are currently under house arrest, and she says every single family member has struggled with some type of addictive behavior, ranging from eating disorders to drug problems. Trinity knows what put them on that path.

When she heard Bruce speak recently about how childhood experience affects the brain, she wanted to share her story, because so much of what he described helped explain the way she saw and experienced the world. We wanted to include it here because it illustrates several key factors in the development of empathy under stressful conditions.

We also wanted to be sure we didn't make the same mistake in this book as Trinity's elementary school teacher did with her class. Although we explore critical factors that can affect the development of empathy—both for good and for ill—throughout these chapters, we know that there's more to every human story. We are far from fully understanding the intricacies of the human mind, our wills, and our behavior. And so we want to stress: increased risk doesn't mean inevitability; disadvantage is not destiny.

Though obviously, it is better for children to be raised in warm, safe, loving homes, those who do not have perfect or even acceptable early parenting are not doomed to be either dysfunctional or dangerous. Kindness produces more kindness, but lack of empathetic parenting, fortunately, is not always a disaster. Indeed, the majority of children of addicts and alcoholics do not become addicts or alcoholics themselves; most abused children don't become abusers. It's certainly true that they are at much higher risk for these and other problems than children who don't have those predisposing experiences and/or genetics. That's part of why we have to improve the situations of these children (the other part, of course, is simple compassion). Nonetheless, predispositions are not predestination.

In fact, paradoxically, being treated cruelly allows some children to become more kind and empathetic, not less. They know what it's like to be hurt—and they want to protect others from such experience. Posttraumatic dysfunction or disability is not the only possible result of trauma—there's also posttraumatic growth. But it occurs only under certain conditions.

TRINITY WALLACE-ELLIS WAS just five years old the first time she was put into foster care. Someone had reported that she and her siblings were being abused and neglected; she doesn't remember much about it. Her parents were sent to parenting classes, her father cleaned up, and the children were returned. But, unfortunately, their home didn't get any safer.

When she was in first grade, her parents split up and her mother found a new partner. Rex did not have any children of his own and was unprepared for life with his new girlfriend's eight young girls. He knew nothing about child development, nothing about what could realistically be expected from infants and toddlers. When child welfare authorities discovered abuse again, the girls were sent back to foster care. But after a short time, they were brought home, this time contingent on Cassandra keeping Rex out of her life. Trinity says she told the caseworkers that he would be back—that they shouldn't believe her mother's promises about men. They ignored her; she was only a child, what did she know about adult behavior? "Slowly but surely, the clothes started building up in the closet, he started spending the night, and then he was there all the time, and he was mean," she says.

Rex beat the girls with rubber strips taken from screen windows and other implements that caused serious pain and bruises. One day, he decided that Cassandra spent too much of her time styling her daughters' hair. When Trinity came home from school, she found her sisters with short, uneven haircuts, looking like "little orphan Annies."

They were forlorn, and she was angry. Trinity refused to let him cut her own hair—but was unable to talk her mother out of doing so. Rex didn't seem to have any concern for how the girls felt. He didn't seem to know or care how important it was for their mother to spend time with them on everyday tasks. Nor did he consider how they might want to choose their own looks, how their hairstyles might be part of how they defined themselves—or how much they would now be teased at school. Trinity couldn't understand why her mother kept putting Rex first, why she didn't see how inhumane he was. She just tried to protect herself and her sisters when she could.

Trinity believes that several factors allowed her to survive to become a compassionate advocate for children. The first was her own empathetic awareness, a sensitivity that was present in Trinity from her earliest memories. "I could feel the pain of the people around me," she says. "From the time I was very young, I was very aware of my surroundings and what was going on. They always said I had wisdom beyond my years." This sensitivity—which certainly has both genetic and environmental influences—allowed her to tune in to the love that her biological father did have for her, underneath all of his dysfunctional and even cruel behavior.

Her earliest memory, in fact, is being in a motel room with her parents. Because she was the only child there and her mother was pregnant, she thinks she must have been about two and a half. Her father was changing his clothes, so he told her to look away. She remembers that he was patient when she didn't comply immediately—which was uncharacteristic for him. And she felt loved when he said her name and calmly repeated himself, as she lay there next to her mom. Trinity also fondly remembers her father calling her his "number one" because she was his oldest girl, his sidekick. She liked helping him count money and feeling useful. Amid all the chaos of his addiction and illegal lifestyle, she was able to find and cherish positive moments. "The love that I got from my father when he was sober is what I looked forward to," she says. "I knew it was a possibility, and I kept that memory fresh in my mind, and kept it fresh in the minds of my sisters, and I just loved them. Knowing that they were looking up to me, it's what made me want to do better." Trinity's father always stressed that his girls should take care of one another. She took that lesson deeply to heart. By mothering her sisters, she helped parent herself as well.

Trinity's sensitivity and ability to empathize are probably related: studies show that people who are more sensitive to the emotions of others are more likely to mirror their facial expressions and body language. Those who mirror more also rate highest on measures of empathy.[1] Her sensitivity was probably also reflected in the responsiveness of her stress system.

When she recently attended a talk that Bruce gave about how children raised in threatening environments are frequently in a hyperaroused state of "alarm," Trinity recognized herself, both as a child and an adult. "He mentioned how when a child is exposed to trauma, it can have effects that can carry with them into adulthood because their body is in a constant state of high alert," she says. Though she has now made a wonderful life for herself, she still carries physical and emotional reminders of her past that affect her mind and body every day.

So why would Trinity have a "tuned-up" stress system—and how could this help her survive a traumatic childhood? Being hypervigilant is self-protective: a child who notices the signals in tone and body language that precede the abuse of an abusive parent can often avoid, evade, or even sometimes defuse the situation. This stress response is adaptive. Although, as we'll see, there can be long-term harmful effects, in the short term, being hyperaroused serves a useful purpose, at least in some settings.

In contrast, reduced stress system responsiveness as measured by levels of cortisol and skin conductance is often linked with antisocial behavior. As noted in Chapter 5, research looking at children of criminals who did *not* follow in their parents' footsteps showed that their stress systems were more likely to be set on "high," like Trinity's— not low, as seen in Danny's family. During her childhood and even into adulthood, Trinity was constantly watchful and observant, considering what other people were feeling and thinking.

Trinity's keen intelligence also gave her an advantage. Research on resilience shows that smarter children are often better able to cope with chaotic and stressful childhoods.[2] Having higher intelligence can allow a child to come up with more alternatives when faced with difficult situations. Oftentimes, the most violent and dysfunctional behavior occurs when people feel trapped or as though they have "no choice." Intelligence allows a child to devise a greater number of creative solutions and to find ways to feel less "cornered." It can also attract the attention of caring adults at school or outside the home. Obviously, as Ryan's story suggests, intelligence isn't universally protective, but even his story shows some resilience due to cognitive skills. If he'd been less bright, he would probably have had earlier and much more visible behavior and academic problems: across the board, he would have been far less functional. While intelligence isn't a complete shield from developmental trauma, studies show it often makes a real difference.[3]

"The best thing my parents taught me was how I didn't want to be when I grew up," Trinity says. "I moved in the opposite direction. I have told many children that I worked with, 'If you don't want to make the same mistakes that your parents made, that your brothers and sisters have made, you need to be aware of where you come from and of the circumstances that have led up to where you are now. I always do this thing in my head, I think about cause and effect, and I understand how I made it here."

Conversely, children who develop severe behavior problems often lack this ability to link cause and effect. The early chaos of their environments doesn't make these connections clear and visible. There are too many anomalies, too many inconsistencies. A child of average or below average intelligence won't be able to learn without consistent repetitions. Trinity's intelligence, on the other hand, meant that she needed fewer repetitions before she saw connections. Her brain cells needed less input to wire in these links. This allowed her to make the critical association between cause and consequence even without much consistency from her parents. Her sensitivity also helped her take the perspective of others, seeing both the pain that her parents' behavior was causing to her and her sisters—and sensing her parents' own pain as well.

But probably the most important factor in Trinity's success was her ability to see the good in others and find caring people outside of her family to help. Her father's behavior was often atrocious—but she focused on the few times he was kind and patient. Most of his friends and customers were hardly models of mercy, either—but she was able to find solace in the occasional kind words and actions of neighbors, teachers, and counselors. This capacity to reach out and attract aid—and the occasional presence of people who did respond with kindness—helped allow her to overcome what many of her siblings could not. The ability to find and connect with nurturing people outside an abusive family is another factor frequently linked with resilience in the research.[4] Trinity's sense of responsibility for her sisters and her relationships with them were also important. All of this helped her develop empathy.

When Trinity was nine, she was placed in therapy by the child welfare system in an attempt to help her cope with having been in foster care and the abuse and neglect that preceded her stay. In the afternoon, her mother would bring her to individual therapy; afterward they would go home for dinner and then return for group. Not surprisingly, the therapy wasn't especially effective. "I felt really immature playing with dolls and stuff, with them waiting to see if I'm going to undress them and lay them on top of each other," she says. She felt like it was "stupid and a waste of my time." Even if she'd had the best therapist in the world, however, it would have been hard for a few hours a week to compensate for what she was seeing and experiencing at home. The sessions became just another responsibility, on top of school and taking care of her sisters. And Trinity had no idea that even worse was to come.

One day, Cassandra was bringing Trinity home between therapy sessions for dinner. They drove up the street and found their house surrounded by ambulances, fire trucks, and police cars. Neighbors were out on their porches and lawns, rubbernecking at the scene. Cassandra asked a neighbor what was going on. "It's one of your babies," the neighbor said. Hearing that, Trinity flew from the car, just in time to see her two-year-old sister being taken out of the house on a stretcher. Inside, she found an older brother on the couch in handcuffs, and she heard neighbors shouting "He went that way" at the police. Her sisters were scattered around, crying. In the kitchen was a half-eaten cake, crumbs spread everywhere.

Trinity began to piece together what had happened. Apparently, two-year-old Righteousness and her four-year-old twin sisters had found and begun to eat a cake that Cassandra had made for Rex. Furious, he had spanked the girls, then dragged Righteousness into the bathtub, holding her head under water. He didn't understand that a two-year-old can't really control her behavior, that children that age need to be supervised if they are to stay out of trouble—or he just didn't care. He didn't let Righteousness breathe until it was too late. She lay still. He fled. Now the police had the wrong suspect on the couch. And the EMTs and soon the emergency room staff were desperately trying to revive Righteousness. Unfortunately, it was too late.

A few hours later, a neighbor told Trinity that her little sister was dead. "If there was ever a way for a nine-year-old to have a nervous breakdown, I think I did. I just went numb and kind of blacked out," she says.

In the face of inescapable trauma, when the brain senses that activating the fight-or-flight system could do harm, it moves in the opposite direction and uses an alternative stress response. This reaction assumes the possibility of injury, slowing blood flow and heart rate, rather than accelerating them. It aims to reduce blood loss if injury is unavoidable. Large amounts of brain opioids are also released, numbing and calming in an attempt to prevent actions that might make things even worse. Although traumatic stress under fight-or-flight conditions can vividly enhance memory, the opposite occurs here. It is often a mercy.

The intense emotional pain and helplessness Trinity felt over her sister's death had produced what we described earlier as a dissociative state. This numb, distanced state is often compared to feeling like you have left your body or you are watching events around you as if in a movie. Memories of such experiences are scattered or even nonexistent, "blacked out." Experiences stressful enough to produce dissociation are likely to cause posttraumatic stress disorder, in which dissociative states and vivid memories from hypervigilant reactions recur, often as intrusive images, thoughts, or nightmares, or much less frequently as a full-blown "flashback" to the original traumatic event.

As we've seen, the stress response system is built to be calmed by social contact, initially through the soothing that comes from being cared for by one's parents as an infant. Normally, this wires the relief of distress to the presence of safe and familiar people. Trinity did have these bonds with her sisters and she clung to the remaining girls—but unfortunately, as they'd done earlier, her parents once again failed her spectacularly.

Trinity's mother became even further unhinged. Not long after her sister's death, Trinity came home from school after her first day of fourth grade. Her mother was moving crazily through the house. Outside, the car was fully packed with her mom's clothes and most cherished possessions. Cassandra knelt down and gave her daughter $55, a promised reward for the eleven A's Trinity had gotten on her previous report card. But then, she said she had to leave. She told Trinity that she loved her, but she had to "go away," to "find herself." Trinity says, "I just remember standing there watching the brake lights come on when she hit the corner, and then the blinker went on and she was gone." She'd left the refrigerator stocked with food—and abandoned her seven little girls who had just lost their youngest sister.

By this time, Trinity was used to coping with extreme circumstances. Little could shock her. She simply did what she had always done and got on with it. In her home, the adults never got up early enough to take the children to school anyway—that had become Trinity's job. So, the next day, without missing a beat, Trinity took her sisters over to their neighbor Myrna's house for breakfast.

Myrna was one of the people who helped preserve Trinity's life and sanity. A kind, maternal woman, she showed Trinity that there was more to the world than violence, abandonment, and neglect. She fed the often-hungry sisters a full breakfast of eggs, OJ, cereal, and toast, similar to the one she prepared for her own diabetic daughter. She also took care of the youngest girls while the older ones were at school. Either Cassandra had paid her in advance or she just didn't ask and continued to babysit as Trinity ran the household. Without Myrna, Trinity's story may well have had a less happy ending. Myrna's actions showed that there were people who cared out there, people who could be relied on to help.

Extreme stress continued to make its mark on her, however. Although she was just nine, Trinity had sped into puberty and now looked a lot older than she was. Early puberty is more common among children living in high-stress circumstances like Trinity's: studies find that having lost a father and/or living in a stressful situation can accelerate the pace of sexual maturation and lower the average age at which people become sexually active.[5]

Indeed, the same types of changes take place in the pups of the less nurturant mother rats discussed in the last chapter—with both earlier sexual maturation and younger initiation of sexual activity.[6] This could help them increase the odds of passing on their genes: if an environment is marked by scarcity and deprivation, long life is unlikely. Preparing for early reproduction may be the only way to ensure that it will happen at all. One study, in fact, found that neighborhoods with the highest murder rates also had the highest teenage pregnancy rates.[7]

Of course, teenage pregnancy in the modern world perpetuates poverty and can keep the next generation tied to stressful neighborhoods and poor environments for child rearing. And early puberty can make girls vulnerable to predators. But Trinity was able to use her physical maturity to her advantage in one way: she was able to pass as her sisters' mother at food banks and other places where she sought aid, keeping food on the table when the adults had failed to do so. When she ran out of resources, she contacted her father. He took custody—but he was still addicted and still abusive. Soon, the girls wound up in foster care, permanently. At first, they were split up and sent to different foster homes.

It was at this time that Trinity began to build the relationships that would help her find her way out. One of the most important of these was with a counselor at school. Donna Brady saw something in Trinity, despite the fact that she was a bit of a class clown. Trinity's intelligence and compassion shined through. A sixth grader now, Trinity was obviously bright and had real academic potential, even though she never did her homework. Brady recognized Trinity's hidden strengths and signed her up as a peer counselor.

"I think that was the defining moment for me," Trinity says. "It showed me that I could make a difference in the lives of people beyond myself and my immediate family." When other children were struggling, if they got into fights or just needed someone to talk to, Brady would send them to Trinity. "I made them feel better and gave them hope and would just kind of be someone who was there for them. I can say that it changed me, it made me popular; the kids were really happy to have someone to talk to."

Trinity's role as peer counselor also allowed her to transform Chris—a boy who was bullying her—into a lifelong friend. For as long as she can remember, Trinity has always been overweight and she has not been able to slim down. Whenever he saw her, Chris would call her "Shamu" or "Free Willy," pointing and laughing, shaming and embarrassing her. But after he broke into the school's computer lab seeking parts for something he was trying to fix, she became his peer counselor. He had already been suspended and was coming close to expulsion: if he wanted to stay in school, he needed to show it.

Behind closed doors, Trinity saw a very different boy. Chris acted tough, like no one meant anything to him. When she got him to talk a little, however, she realized that he actually cared a great deal about his new nephew, recently born to his fourteen-year-old sister. "I learned that moment you have to find something that they care about before they'll open up, before they'll start thinking that things will be okay."

In front of other people, on the street and in school, however,

Chris continued the bully act. It really got to Trinity. She couldn't understand why he didn't recognize how it affected her. She thought about fighting back. But that's the thing about cruel treatment: although many people think that giving bullies a taste of their own medicine will create empathy, if there is no connection to begin with or no ability to make one, none will be formed by being unkind back. Many bullies are already intimately familiar with being on the other end of violence and aggression. Often, in fact, they are being humiliated or even beaten at home. Being further victimized and punished makes them angrier, not kinder. Fighting back may move them on to a new victim, but it doesn't get them to rethink their behavior. As Trinity's own experience shows, suffering can increase empathy in those so inclined; but it can't do so, otherwise. Unfortunately, many of our attempts to "reform" troubled teens—like juvenile detention, boot camps, and many wilderness programs—do not recognize that being tough on tough children is not the way to make them kinder.

In her counseling sessions, however, Trinity saw Chris's vulnerability, and he came to recognize and rely on her advice and support. She saw that they'd made a connection. And so finally, one day on the street, rather than walking away in shame, Trinity stood up for herself. She turned around and told him, "I'm not putting up with your shit no more. I'm going to kick your ass today." His friends drew in their breath. "Oooooh," they said. He'd been punked by a girl. Chris looked at her, saw that he was in danger of losing a friendship that he actually valued—and backed down. He had begun to empathize with her. She wouldn't really have been able to hurt him physically, of course—but he knew that losing her would hurt terribly emotionally. He cared. And to this day, Trinity and Chris remain close friends, frequently visiting with each other's families and children.

With her self-worth affirmed and strengthened in friendships, by her school counselor and through her role as a peer counselor, Trinity began to see a world outside her family, one where cruelty could be transformed into friendship, where people didn't always let you down. In her foster home, she remained a "little mother" to the sisters she'd been reunited with and other foster children, doing their hair, feeding them dinner, and doing most of the menial work involved in raising them. Her foster parents did their best—but at age sixty, with a house full of children, the foster mother had a hard time keeping up. Still, Trinity could see a way out—and examples of a better life ahead.

Of course, Trinity's road out of poverty and neglect was not without setbacks. She did become pregnant at sixteen, by a drug dealer who charmed her and made her feel special. At night or after school, when she was with him, he was taking care of her—she wasn't babysitting or trying to solve problems for her sisters. It was an escape. But it didn't last long. She found out that he'd gotten another girl pregnant during their relationship, and she moved on, devoting herself to her baby daughter. During the pregnancy, she bonded with her foster mother who had no biological children of her own and continues to be a part of Trinity's daughter's life.

And, through an outreach program to young mothers, she was introduced to an agency that organizes foster care youth to advocate for themselves. California Youth Connection soon had Trinity testifying before the state legislature, pushing for a law to require that state child welfare authorities prioritize keeping siblings together in foster care. If a placement in the same home was not possible, the law sought to keep siblings in the same neighborhood and to have caseworkers ensure that the children could stay in contact with one another. Trinity had had the experience of being separated from her sisters and brothers and spoke eloquently about how hard it was and why it mattered to keep families connected. She was given the chance to choose the number that would label the bill, deciding on 1987 in memory of Righteousness, who had been killed that year.

At sixteen, Trinity fully expected to remain a single mother: that was what she saw all around her, and all of her sisters have at one time or another had abusive relationships with men. She didn't believe she would ever have a lasting and healthy relationship. But when her daughter was about a month old, one of her younger sister's friends insisted that his older brother's friend would be a perfect match. Trinity was resistant. "I was like, 'OK. Well, he can't do drugs and he can't be just hanging out—does he go to school?'" But the boy maintained that the guy he had in mind met her criteria and more. He didn't give up. Trinity eventually relented and agreed that he could give him her number.

And sure enough, nearly a month of phone conversations yielded a real sense of connection. Before their first date, Trinity was getting ready and her sisters gathered at the window to check out her prospective beau. One said, "He looks strong, he looks okay!" Another said, "Kinda big, he's cute." The consensus: "You'll like him!" Taj turned out to be kind and hardworking and supportive—and her sisters, and eventually even her foster mom who was worried that he was several years older, approved. He made her feel safe. After their first date, she knew they'd stay together—and they have been, ever since.

Despite a series of illnesses that hospitalized her daughter repeatedly during Trinity's senior year of high school, she graduated and went on to college. Criminal proceedings related to her stepfather's long overdue arrest for her sister's murder interrupted her college years, but she has since gone on to high-level professional jobs in child advocacy. Over the years she has worked one-on-one with over 1,500 children whose own resilience has become, in part, a reflection of hers. Her deep empathy has helped grow that quality not only among her relatives and children—but in nearly everyone who meets her. To talk with her is to know grace, compassion, and mercy are at work in the world.

THERE'S ONE IMPORTANT aspect of empathy, however, that has sometimes eluded Trinity. It's not about empathy for other people: there,
she's a beacon of hope that few others could equal, let alone sur-
pass. However, forgiving herself, allowing herself to be human, and
believing that she deserves happiness has been much more difficult. She's particularly hard on herself regarding her weight. Trinity
currently weighs over three hundred pounds and suffers from high
blood pressure. She experiences cycles of binge eating that she has
not been able to keep under control. She knows this isn't healthy. It
worries her family and that worries her. "I just struggle with it," she
says. "I don't know what it's going to take for me to be able to conquer
this, and I don't want to die before I'm able to do something about it or
get myself in a better frame of mind. And I dread going to the doctor
because he's going to blame everything on my weight, not knowing
my whole history."

Until she heard Bruce speak, Trinity didn't fully realize just how
much her weight problem is related to her childhood. She kept blaming herself. However, a groundbreaking study of more than seventeen
thousand Californians enrolled in the Kaiser Permanente health plan
has shown that childhood trauma is a critically overlooked factor in
the obesity epidemic—and in virtually every other major cause of death
studied. The risks for heart disease, stroke, depression, diabetes,
asthma, and even many cancers are all affected by trauma-related
changes in the stress response system. Empathy and connection affect physical—not just mental—wellness and health.

Trinity's journey perfectly illustrates the pathway to obesity that
Vincent Felitti discovered when he started that study in San Diego. It
became known as the Adverse Childhood Experiences (ACE) Study.
It all began with a curious problem that Felitti kept having as he ran a

weight loss program. Felitti, who comes from a family of doctors, is the founder of Kaiser's Department of Preventative Medicine. In the early 1980s, he recognized the need to fight obesity and began developing a program. He soon found that many obese people could successfully lose large amounts of weight if supervised on the supplement Optifast. But for some reason, his best patients kept dropping out.

"This is pretty bizarre behavior," recalls Felitti. "Here we are helping people do what they say they want to do. Somebody comes in with two hundred pounds to lose and they lose seventy-five and then flee the program without saying good-bye. That was pretty annoying." Felitti couldn't figure out what was wrong. He could understand why people who thought they were failing would drop out. But why would those who were doing well quit? He began to ask patients about when they first gained weight and what they were feeling when they started to lose it.

And, after surveying nearly three hundred morbidly obese patients about their life experiences, he was struck by the extraordinarily high prevalence of severe family dysfunction, particularly sexual abuse. About 50 percent of his patients—male and female—reported having been sexually assaulted as children. That is more than 50 percent higher than for the general female population and more than triple the rate for men.[8] Virtually all of the rest of his subjects had experienced some lasting form of childhood trauma.

Felitti recalls presenting these findings at a meeting of a major academic obesity society in 1990. He expected to generate interest and discussion. Instead, he was attacked and dismissed. One audience member even stood up and claimed that the patients were merely making up stories to cover up for "failed lives." But David Williamson, a researcher at the Centers for Disease Control and Prevention (CDC), was intrigued. He approached Felitti and said that if what his research suggested was really true, it had major implications for public health. However, a large epidemiological survey would be needed to prove it.

Felitti realized almost immediately that he was perfectly positioned to do such a study. Kaiser Permanente conducts medical exams on more than fifty-eight thousand people each year. If even a small portion agreed to be surveyed about their childhoods and to allow the anonymous use of their medical records, that would be a sample large enough to use to draw convincing statistical conclusions. Felitti might be able to discern previously unrecognized connections between childhood histories and adult health, even if they were quite subtle, with a sample that size.

And so, the ACE Study was born. Another researcher at the CDC, Robert Anda, joined Felitti, becoming coprincipal investigator and conducting much of the epidemiological analysis. "Adverse childhood experiences" were defined as the following: ongoing emotional abuse, ongoing physical abuse, sexual abuse involving physical contact, having one or no parents, living with someone with an alcohol or other drug problem, living with a mentally ill family member, losing a household member to some period of incarceration, and domestic violence against one's mother. A second wave of the study also included emotional or physical neglect as an ACE.

To track the cumulative impact, each one of these experiences is counted as one ACE. As the number of ACEs increases, the risk for both mental and physical illness rises in concert. Having four or more increases risks for some conditions as dramatically as smoking raises the risk of lung cancer. Although some ACEs may also reflect genetic risk (e.g., having an addicted or mentally ill parent), overall, they give excellent insight into a person's childhood environment. (If you want to figure out your own ACE score, the website, acestudy.org, has a questionnaire/calculator.)

The relationships between ACEs and adult health problems are stronger than for many of the typical health risks authorities are currently fighting, like smoking. Compared with having none, having four or more ACEs increases the risk of ischemic heart disease by 220 percent, stroke by 240 percent, diabetes by 160 percent, and chronic bronchitis or emphysema by 390 percent.[9] Having an ACE score of six or more can lower life expectancy by nearly two decades.[10]

So how does childhood trauma have such a massive impact on physical health? As you may suspect by now, it's related to the stress response system, which affects every aspect of mental and physical functioning. This system influences the heart, brain, blood vessels, immune system, and digestive tract: it reaches every single cell and organ. Its hormones are known to have damaging effects on the brain and body if they remain active for too long. The cascade of emotional, social, and physical health vulnerabilities that flow from a dysfunctional stress response system is profound.

Another part of the story is the resultant sadness and depression: having four or more ACEs increases the odds of past-year depression by 460 percent and the chances of ever attempting suicide by a massive 1,220 percent.[11] Trinity knows all too well the deep sadness that can still overtake her at times. And of course, a further result of trauma is the way people attempt to deal with related pain by self-medication or self-soothing with food, cigarettes, alcohol, and illegal drugs. (Food was Trinity's substance of choice.) Having four or more ACEs increases the odds of being a current smoker by 220 percent. It raises the odds of ever using illegal drugs by 470 percent. The chances of becoming an alcoholic go up by 740 percent—and the odds of injecting drugs climb by a whopping 1,030 percent.[12] "I remember people at CDC doing this work, and they told me that those numbers were numbers the magnitude of which epidemiologists might see once in a career," says Felitti.

But even those who don't misuse alcohol or other drugs or become obese have measurable elevations in their risk for high blood pressure, heart disease, stroke, and some cancers and infectious diseases.

This, as we shall explore in more depth in Chapter 11, is due to stress
system adaptations, which are built to manage short-term extreme
stress but can damage the brain and body if severe stress becomes
chronic. It's awful to think that being traumatized as a child could
have so many long-lasting negative effects. However, understanding
their impact could lead not only to more effective treatments but
also, we hope, to greater determination on the part of parents, policy-
makers, and politicians to prevent those traumas that can be avoided.

Such understanding can also help people like Trinity, even before
further discoveries are made. When Maia told Trinity about Felitti's
work with obesity, she cried. She said that she'd discovered very early
in life that if she was fat, her father's friends and customers would
leave her alone. She says, "I could hide behind my weight and then
they wouldn't want to mess with me. I felt guilty for being the fat sister
because I didn't have all the sexual abuse that [my other sisters did].
They were the skinnier ones and they were the prettier ones, and I
used to say, 'The bigger I got, the more they would leave me alone.'"
Or, as one of Felitti's patients put it, "Overweight is overlooked, and I
needed to be overlooked."

Felitti currently runs support groups for his weight loss patients to
help prevent the attrition that started him on this journey. He says,
"Ultimately we came to realize that obesity is not the problem. Obesity is the marker for the problem, somewhat like smoke in house
fires, and indeed, sometimes, it's the solution." In his introductory
meeting for participants, Felitti asks them first about why they think
people become fat and why they tend to regain the weight. Most people
converge on the notion that there are psychological reasons for both—
and that these must be dealt with in order for weight loss to be maintained.

Then Felitti says, "'Tell me some benefits of being fat.' That's an
ugly question. And you can see them pulling back in their chairs, their

arms over their bellies. But people will talk about it being sexually protective, they will talk about it being physically protective and socially protective, that people will expect less of you." He describes several prison guards who felt too vulnerable and stressed to handle their jobs if they weren't bulked up by fat as well as muscle.

Felitti believes that the current obesity epidemic has multiple causes. He certainly isn't claiming that there was a recent, sudden outbreak of traumatic childhood experience. Decreased daily activity and the increased availability of high-fat and high-sugar foods—many of the factors that have been widely discussed—have certainly played their part. However, he has found that among the most obese patients, the onset of weight gain is sudden: people don't gain two hundred pounds ten pounds at a time over the years. "The single most important question to ask people in terms of obesity is how old were you when you first began putting on weight," he says. The gain often occurred immediately after a disturbing life event. Felitti identifies parental divorce as an unrecognized factor in the fattening of America and a common precursor to weight gain. Divorce rates did rise dramatically just before obesity began to spike in the United States.

In Trinity's case, her intensely chaotic early life probably set her up for addictive behavior and high blood pressure even before she could consciously make choices about food. A further precipitating event was the sexual abuse and discovering that being overweight would stop her from being victimized. She recalls that after her father had beaten her, the next day she'd often find a spread of Entenmann's desserts waiting for her when she came home from school. For her, food was an especially strong source of comfort. Realizing these connections—and the way her weight is linked directly to the childhood stress she experienced in a way that is beyond her conscious control—has been liberating.

"When I first heard of Bruce's work, I thought, well, maybe some of this isn't my fault. Maybe it's not all my fault," she says, considering the idea again.

As we saw in the last chapter, Ryan's response to early loss of important relationships was to keep all future relationships shallow, engaging only to bend people to his will. Not finding much pleasure in connection, he sought it elsewhere. But Trinity took the opposite tack: she has filled her life with so many relationships and connects so deeply with people that any diminishment of relational pleasure that her early experience may have produced is compensated for by this abundance. She says she copes with the pain that remains by recognizing the positive effects she has on others.

"I've been very fortunate to be able to see that I'm making a difference," she says. "So that when things kind of suck at home, I'm able to see it at work. When things aren't going right at work I can see it at home, or in some other volunteer work that I'm doing. And that's what I need to see."

She adds, "When things are out of my control is when it gets hard. That's when I'm more susceptible to depression and binge eating, but I've been pretty good at being able to identify my cycles and kind of be aware of what's going on with myself. Even when I am feeling sad or depressed, I'll get a phone call or somebody will ask me to do something that takes the focus off of me and I'm back at it." She tells the youths she works with, "I don't promise that it will be easy. I do promise them that they can get through it, as long as they keep putting one foot in front of the other and getting up every day. That's what I do, I just keep getting up."

TRINITY'S RECOGNITION OF the way her early life has affected her is an ongoing part of her resilience. She uses her insight to keep on going, to inspire others. From her story, we can see how we hold much of our children's health in our hands. Our capacity for caring—and lack of care—affects both physical and mental health across the life span. But although our individual and family relationships set the stage, as children grow, their relationships with peers and their ability to negotiate larger social groups take on increasingly greater importance. For Trinity, she found hope in those groups and organizations outside of her family. In the next few chapters, we'll look more closely at how empathy affects—and is affected by—specific group pressures, in both positive and negative ways.

S HE WAS ONLY TWELVE BUT SHE looked—or was trying to look— like a teenager. She was dressed entirely in black, her hair dyed deep ebony with a streak of red not found in nature. Her fingernails were sloppily painted black, too. She had a nose ring, multiple ear piercings, an eyebrow piercing, and a few small tattoos. When Bruce walked into her hospital room, Alyson was thrashing around on the bed. Her fingers were covered with multiple rings, including one that was shaped like a skull.

Tall for her age and pale, Alyson seemed to be trying to show the doctors and nurses that she was very seriously affected by the over-dose of Tylenol that she had taken. But from her chart, Bruce knew that she hadn't ingested a quantity sufficient to damage her liver. That was lucky, because even though it is sold over the counter, the active ingredient in Tylenol (acetaminophen) can be fatal in high doses. The hospital staff had pumped her stomach with charcoal just in case. This practice was also an attempt to make the experience sufficiently awful so that youths like Alyson would tell their friends how bad it was and think twice before trying such a stunt again.

Alyson was a straight-edge Goth: when they talked, she informed Bruce that this meant she opposed any and all psychoactive drug use, including alcohol. But her boyfriend—who was similarly inclined—had recently dumped her. To add insult to injury, he was now seeing a girl she knew. Without him, Alyson had decided that life just wasn't worth living. Not, however, before calling him and her closest friends to detail how hurt she was and that she was dying. Freaked out by what Alyson said on the phone, a girlfriend called her parents and now here she was.

Bruce knew that her flailing was an obvious attempt to act drugged, even though Tylenol does not alter consciousness. This was a cry for help, rather than a serious suicide attempt. He asked if she'd be willing to meet with him to talk about what happened and about why she felt so desperate and miserable. They set up an appointment, and over the next few months, Bruce began to put together a history of a highly sensitive girl who desperately wanted to fit in. She wanted to be part of a group and would do whatever it took to do so. Alyson was a chameleon.

Throughout her childhood, her parents had moved every few years to accommodate her father's high-powered job in the oil industry. At the start of elementary school, she was a smart, kind, popular girl—getting good grades and making friends easily. Aside from her academic ability, nothing stood out. At that age, doing well wasn't particularly stigmatized as "nerdy" or "geeky." When she entered fifth grade, however, her father was sent to Venezuela—and the whole family moved along with him. Her parents enrolled the ten-year-old in a school for the children of ex-pats and diplomats. But because she was one of only a few younger children among seventh and eighth graders, making friends was a lot harder for her.

Alyson adapted rapidly, however. Though she hadn't yet hit puberty and wasn't really interested in boys, she began dressing provocatively. The other girls were obsessed with fashion and music—so she was, too. They wore makeup, low-cut tops, and short skirts—and so did she. When they snuck out at night to meet boys, she did like-wise, although she didn't actually do more than kiss them. Her parents became concerned and tried to rein her in, but sometimes she eluded them. When they moved back to the States the next year, Alyson thought she'd have it made. She'd been in with the cool kids, even though she was several years younger. She was sophisticated, had seen the world.

But in her new school, Alyson struggled. The kids there weren't like the ex-pats—all they wanted to do was play sports. They looked at her like she might be a slut and thought her talk about how great it was in other countries was un-American and snotty. OK, she told herself, if sports is what they like, sports is what I'll give them. Soon, the tousled hair and short skirts were replaced by jogging shorts, tank tops, and a ponytail. Rapidly, the volleyball team became the center of her life. As enthusiastic as she'd once been about her "mature" image, her energy was now directed at being a jock. She now ate, dreamt, and slept volleyball. She loved her teammates and would do anything for them. Once again, she'd changed her colors to blend into a crowd.

MANY CHILDREN GO through phases, of course—and virtually every teenager works hard to fit in with friends. During adolescence, teens who don't try to create a social life for themselves are the abnormal ones: this is not a sign of health at this age. But why do friendships become so important—particularly for girls—as they hit puberty? Why do teens and tweens work so hard to fit in? And what can this teach us about empathy and the way it can disappear so completely when human groups conflict with one another?

Alyson's chameleon-like behavior showed that she had developed a keen eye for detail and an ability to pick up which aspects of her friends' styles were best to appropriate. Her emotional sensitivity allowed her to figure out exactly which aspects to copy to belong. As we've seen, for most people, the basic ability to imitate the behavior and emotions of others starts in the cradle. Even newborns can mimic some facial expressions. Like dogs and cats, most babies are easily affected by the moods and emotions of people around them. This simple "emotional contagion" is one basis for the development of empathy.

But it takes on a new character for adolescents, whose developmental challenge is to start to master the hierarchies, politics, and culture of human groups. The longing for just the right amount of conformity—the desire to fit in above all but stand out just enough to get recognized—is pervasive. This yearning becomes increasingly compelling as tweens become teens.

The vulnerability of young people to emotional contagion is well known: American culture is marinated in it, as trends in music, fashion, TV, movies, and other media are picked up and spread from one teen to another, hyped by an unprecedented global marketing machine. The popularity of the biggest celebrities relies heavily on their ability to attract adolescent interest and then capitalize on the copycatting that results from teen longing to belong. A more extreme manifestation is teen susceptibility to what is now called "mass psychogenic illness" (formerly, and less politically correctly, epidemic hysteria). In these cases, a few people suddenly develop unexplained symptoms— for example, epileptic fits, breaking out in hysterical laughter, or vomiting—which soon spread to others nearby. The most common victims are teen and preteen girls.

One suspected historical case was the outbreak that produced the Salem witch trials of the 1600s. There, a teenage girl and two preteens began having what seemed like severe seizures, which were interpreted as being caused by spells cast by witches. The terrible outcome is well known, but similar cases in which groups of people develop similar medically inexplicable conditions after exposure to one another continue to occur. In 2007, for example, a journal article described a series of cases in which fourteen teen girls attending high school in Florida were suddenly afflicted with wheezing and strange high-pitched sounds upon intake of breath. No medical cause could be found and the illness eventually passed.[1] Other cases have involved dozens of teens and young adults, all suddenly overcome with similar, mysterious symptoms that fade with time and distance. Teens' increased sensitivity to social signals probably explains their increased susceptibility to both fads and these strange epidemics.

This is another case in which our ability to mirror and take the perspective of others isn't always a positive thing. Perspective taking, though a component of empathy, can also be used to manipulate people, to find their weak spots and exploit them. Its physiological basis in mirroring can also make us vulnerable to group pressures to conform. After all, mirroring others comes naturally as we pick up their perspectives—this allows not only one-to-one connections, but group solidarity. However, peer pressure often includes a simultaneous drive to exclude and reject others. As does popularity: research finds that those who climb to the top of the social ladder are those who can best balance exclusivity and kindness toward others.[2]

Further, one crucial factor that determines how much we empathize with someone is whether or not we feel we belong to the same social group. There are specific networks in the brain devoted to determining whether an individual is one of "us" or one of "them"—and if someone is categorized as "them," the facility for empathy can be deeply reduced, even shut down entirely. From early clan warfare to the Nazis and Rwanda, human history is sadly replete with examples of devoted loyalty to one's "in group," and the most brutal and indecent inhumanity imposed on those outside of that charmed circle. The system that allows "us" to target "them" comes into sharp relief—and can be at its most dangerous—during adolescence and early adulthood.

This was certainly true in Alyson's case. Her suicide attempt would not be the most dangerous behavior she engaged in—far from it. As we shall see, her desire to fit in began to put her—as well as those around her—at great risk. But no one—including Bruce—saw it coming.

The key to Alyson's story lies in the roots of our "groupish" nature. As we've seen, some researchers have argued that the evolution of the human ability to cooperate was driven primarily by the need to compete with other groups. This explanation goes all the way back to Darwin, who wrote that groups with "a greater number of courageous, sympathetic and faithful members who were always ready to warn each other of danger, to aid and defend each other . . . would spread and be victorious over other tribes."[3] This would increase the odds that genes that promote cooperation would survive—despite the fact that for individuals, cheating to get the benefits of cooperation without the risks could often promote survival. If groups with more cheaters battled groups of altruists, the altruists, with their greater participation in group efforts, would tend to prevail.

A study published recently in *Science* used archaeological evidence, research on modern hunter-gatherers, and mathematical modeling to explore whether warfare was common enough or deadly enough to actually give altruistic groups a survival edge. All three lines of evidence converged to support this theory. Given how lethal early human intergroup conflicts appear to have been, genes that promoted altruistic behavior would definitely have been favored in that kind of environment.[4] A tribe made up of heroic warriors ready to die for their brothers would certainly have prevailed over one in which disloyal cowards scattered and betrayed one another. If history is the tale told by the victors, this legacy is echoed even in the tone of the language available to contrast such groups.

But while conflict between groups can account for why altruism would persist in humans and how cooperation could become highly developed, it may not be able to explain why it would exist in the first place. Though there is a great deal of evidence for early human warfare starting about ten thousand to fifteen thousand years ago, so far there is no evidence of war in the Pleistocene, the period when early humans and their closest ancestors first appeared. This doesn't necessarily mean that wars didn't take place. Nonetheless, it's quite striking.

This fact and several others have led anthropologist Sarah Hrdy and her colleagues to devise a new theory about why we cooperate. Hrdy says, in essence, that an early human version of "daycare"—not warfare—drove the rise of human empathy and altruism. And, curiously enough, this may help explain why teenage girls like Alyson are so obsessed with fitting in and forming cliques that they sometimes engage in completely uncharacteristic behavior.

Hrdy notes the dismal infant mortality rates that are seen even to this day in countries where modern maternity and infant care aren't available. Angola currently holds the unfortunate record for having the world's highest rate, with a staggering 18 percent of babies dying before reaching their first birthday.[5] In prehumans, Hrdy estimates that half of all children died before reaching puberty. Among the hunter-gatherer Mandinka tribe who was studied between 1950 and 1980, nearly 40 percent of all children were dead by age five.[6]

With this in mind, our species could probably not have survived at all—let alone in numbers large enough to fight wars with one another—if serious energy and attention wasn't devoted to child rearing. Although researchers have typically looked at our closest surviving ancestors, the great apes, for hints of what early human child-rearing styles were like, Hrdy thinks this is a mistake. In great apes like chimpanzees and orangutans, infants are nurtured exclusively by their mothers, nursing from four to seven full years. In infancy, these little apes are in constant skin-to-skin contact with their mothers, day and night. But Hrdy thinks this breeding style of intensely possessive motherhood couldn't have worked for early humans.

Instead, she says, there's good evidence that human mothers with helpers—older children, grandparents, siblings, cousins—were more likely to have surviving children. Certainly, human mothers hold, cuddle, and have a great deal of physical contact with their babies. But unlike gorillas or chimpanzees, human mothers are happy to let others carry, play with, and care for their babies. Typically, a human birth prompts celebration. In all cultures, a welcome new baby is eagerly passed around among the waiting relatives. But among the great apes, a new baby that was passed around would soon be dead meat—literally.

The importance of helpers in human childcare was obscured by several things. For one, the people originally seen as the best model for early child-rearing practices turned out instead to be rather unusual. The !Kung hunter-gatherers seemed to closely parallel the great-ape style of exclusive maternal nurturing. But as it turns out, among all surviving hunter-gatherers, !Kung babies probably spend the most time in physical contact with their mothers. They're at the extreme end of mother nurture—not in the middle. Seventy-five percent of the time, !Kung infants are either in a sling that allows them to nurse whenever they choose, or they are strapped to their mothers' backs. Babies in other hunter-gatherer groups, and certainly modern babies, spend far less time in this way. Nonetheless, researchers had previously ignored the fact that in the great apes, moms and babies touch 100 percent of the time while even among the !Kung, babies spend a full quarter of their time being held by others.[7]

In contrast to humans, if a chimp takes a mother's newborn, that baby is in great danger. Both males and even females can and will kill unrelated babies. Consequently, great ape mothers zealously guard their infants, rarely permitting even close relatives to pick them up.

Indeed, chimpanzees will simply not allow anyone else to hold little ones before they are three and a half months old, and orangutans don't voluntarily allow others any access until five months. Even after that, they only grudgingly let relatives or higher-ranking animals close.

But human moms are happy to let mothers, fathers, sisters, brothers, grandparents, aunts, cousins, and friends hold their infants. We delight in it. The human response to picking up virtually any new baby is overwhelmingly tender. Though some are indifferent, rare is the human who, when handed a baby, will randomly harm her. Most human child abuse occurs among isolated, stressed caregivers—not among those with many helpers. The ubiquity of the kind reaction suggests that early humans must have frequently cared for one another's children. That means that, unlike great ape mothers, human moms must have spent significant time out of skin contact with their infants, though they were likely nearby.

And, as it turns out, the presence of willing "babysitters" may have been even more important to early human survival than the closeness of a father. Among the Mandinka, for example, the research found that having a maternal grandmother or older sister around cut in half a child's risk of dying young. But a father's presence made no survival difference at all—and a stepfather actually increased the child's mortality risk. Most interestingly, once a child was two, the death of her own mother wasn't a factor in the child's survival, either. In fact, orphans got extra attention from extended family, leading to a local expression that translates as "fat as an orphan," which sounds peculiar in any other context.[8]

Consequently, Hrdy argues that although fathers and nuclear families certainly are critical for many aspects of life and health, the role of extended family and friends in keeping children healthy has been overlooked. Cross-culturally, studies have found that poor mothers, single moms, teen mothers, and mothers of premature babies have children who do significantly better on all measures—academic, emotional, and physical—if they have extended family, particularly maternal family, nearby and participating in their lives.[9] Like marmosets and some other small monkeys—but unlike the great apes—humans seem to be cooperative breeders.

Unfortunately, today we seem to be ignoring the important role of our extended families, placing intense pressure on mothers and the nuclear family. Rather than seeing shared care as "natural," we harshly judge mothers who don't spend all of their time with their infants, and often view daycare as a modern evil. In policy debates, we ignore the help historically given to mothers by relatives and friends, viewing motherhood in an isolated suburban home with a father who is absent for most of the day as "traditional." We don't consider the fact that for most of human history, mothers worked with other mothers, with their children around them. Intense attention has been given to the problems of single parenthood and divorce—yet virtually no one looks at the effects of the massive breakdown of extended family that has come with industrial and postindustrial society. When extended family and related social networks are studied, however, they show great benefits to children—especially for single parents, but for two-parent families as well. And, given what we know about how we physiologically need each other to cope with stress, this is far from surprising.

Hrdy argues that cooperative breeding was a driving force in human evolution—and also that it explains a lot about teen female behavior. Cooperative breeders share food and mother's helpers often "babysit," experiences that could promote the survival of genes that increase helping across the board. A species in which cooperative breeding was important to survival would also tend to produce babies who were increasingly good at attracting helpers. Increased cuteness would be one possible result. But a more important outcome would be enhanced survival of babies who are sensitive to emotional contagion and good at reading people. As these genes got selected, people in the groups that had them would become better and better at understanding and caring for one another—producing a virtuous spiral of escalating altruism. Or at least, escalating altruism toward group members.

The same genes might later have allowed cooperation in warfare, which would further select for them—but they probably originated in the earliest versions of babysitting and daycare. If Hrdy is right, women and babies drove the evolution of the caring brain, not men at war. "Were it not for the peculiar combination of empathy and mind reading," she writes, "we would not have evolved to be humans at all."[10]

SO WHAT DOES all this have to do with teenage girls like Alyson? If having female relatives and friends is necessary for the survival of your offspring, being able to recruit such helpers is literally a matter of life or death. As girls become ready for reproduction, building a network of female supporters would then become almost as important as finding the right mate. From this perspective, the dramatic teenage world of cliques and BFFs, queen bees and wannabes makes a lot of sense. Girls would need to learn to cooperate in small groups to achieve success as mothers. This cooperation wouldn't take place automatically—like other skills, it would take practice and would involve inevitable mistakes and misunderstandings.

Alyson's desperate longing for close female friends reflected this. Bruce began seeing her as a patient over the next several years. He learned that her Goth phase had been preceded by an attempt to be a jock at her new school in Texas—but unfortunately, the volleyball skills that had served her well at her previous school hadn't been good enough to win her a place on the team at this much larger school. She talked about how she felt empty and lonely and couldn't really tell where her disguises ended and her real self began. Although Alyson was no longer suicidal, she was drifting. And, over time, her Goth clothes began to be replaced by a new, tougher style. The flamboyant red streak was gone, the black hair was now cut short, and her black outfits had been replaced by tight, designer jeans, heavy biker-style belts, and skimpier tops.

One day, about two years after Alyson's suicide attempt, her parents called Bruce's office, distraught. They were stunned and panicked. They asked if he knew any good attorneys, asked if he'd had any idea what she was up to. They just didn't understand it, couldn't get their heads around what the police were telling them. Alyson had been arrested for a crime that shocked everyone.

Bruce expected to hear something about teen drinking, maybe drug possession, considering that she'd dropped the straight-edge group and, probably, its strict abstinence ideology. But what he was told was far more disturbing: Alyson had taken part in a vicious assault on another girl. The victim's injuries had been potentially life-threatening. The police regarded the incident as a gang-related fight. None of the adults in Alyson's life had had any idea that the friends who'd replaced her Goth crowd were wannabe gang members who had modeled themselves on a seriously violent local gang.

Bruce was taken by surprise; he just hadn't seen anything like this coming. He knew Alyson as a kind, thoughtful girl—a little lost, often dramatic and over the top in her behavior—but in no way given to violence. He went over his notes, looking for signs that might have pointed toward this kind of trouble. But he also knew that the acute desire to fit in with a group can prompt extreme behavior that someone like Alyson on her own would never even consider—let alone do. In retrospect, it was clear that the girl who could blend in with and almost out-Goth the Goths and, at other times, connect strongly with the jocks and young fashionistas would be at high risk if she wound up in a violent crowd.

Research plentifully demonstrates these troublesome tendencies. The best-known studies were done by Swarthmore social psychology professor Solomon Asch and Yale psychologist Stanley Milgram, who was inspired by Asch's work. Asch placed each subject in a group of six to eight people, purportedly to test their vision. They were shown a chart picturing several lines of unequal lengths and asked to find the closest match with a line on another chart. All of the others in the group were experimenters posing as subjects and they consecutively gave their opinions about the best match, the real subject usually being the last one who was asked to comment.

When the experimenters unanimously gave obviously wrong answers, about one-third of the time, subjects did, too. Seventy-five percent of people gave an obviously wrong answer at least once. Interviewed afterward, these subjects said that they'd known that their answer was incorrect, but they were afraid of seeming weird or peculiar. They didn't want to stand out—so they buried the truth.[11] However, some went further, saying that they actually thought they'd chosen the correct item. The perspective of their fellow group members had actually altered the subjects' perception of reality itself! Interestingly, if only one of the experimenters dissented and gave a correct answer, the proportion of answers on which people conformed fell to 2.5 percent.[12] In control groups, virtually no one got the wrong answer on the task because it was so easy, which suggested that the errors had to be linked to social pressure.

Milgram's now-infamous experiments were much more disturbing. As you may recall from psychology classes, they showed that most people would indeed "just follow orders" if directed to hurt another human being. Subjects were told that the researchers were studying learning and that some people would be assigned to be "learners" and others "teachers." In reality, all of the learners were confederates of the researchers. Isolated in a room with a researcher, each "teacher" was placed in front of an instrument panel and instructed to give increasingly strong electric shocks to a single learner when he made mistakes or failed to answer. The learner was located behind a glass window, wired to the shock machine, visible to the teacher and audible over a sound system. Once the shocks reached a certain voltage, the learner started complaining of heart problems. When one of the most painful shocks was applied, he appeared to fall down and become silent.

Dressed in a white lab coat, the researcher stood next to the teacher and encouraged him to continue, even if he began to worry about the learner's health (the first subjects were all men, but the experiment was later repeated with the same results in women). A horrifying 65 percent of participants gave all of the shocks as instructed, even hitting the lever labeled 450 volts, "Danger: Severe Shock," after the learner had stopped responding and had presumably passed out or worse. Most participants didn't do this without questioning, but, pushed by researchers saying, "You have no other choice, you must go on," most did continue to inflict shocks even after the learner appeared injured. These subjects did not try to leave or simply refuse— even though they had been told at the outset that their participation was completely voluntary.[13] Sadly, even the minority who quit the experiment and refused to give the higher shock levels didn't go to check on the "victim" or complain to the university about what they must have thought were serious abuses.

However, if the experiment was done by placing the subject in a group of "teachers"—two of whom were really confederates of the researchers—when the confederates refused to comply, the subject only gave the shocks 10 percent of the time.[14] It was not only the authority of the experimenter that determined compliance—it was also the behavior of the subject's "peers."

Milgram's original research was done in the 1950s. Some thought it worked only because conformity was so highly valued in America at that time. Unfortunately, documentary makers in the United States and Britain have since replicated the results in the 2000s—with some modifications, but the same depressing outcomes.

Given the ethical controversies over doing this type of research on adults, it has never been tried with children or teenagers. But anyone who has ever been or raised a teenager will likely predict even more conformity—and less resistance to social pressure to do harm— among adolescents. Virtually all of us can recall doing something with a group as teenagers that we never would have done alone—or having friends who did so. Consequently, many parents lie awake late at night, waiting for their teens to come home, haunted by visions of what they got themselves up to as teenagers, now realizing how stupid it was to accept the dare to jump off the cliff into the lake or to start smoking because your best friend looked so cool doing it.

Alyson's parents had received one of the many phone calls that all parents dread. Looking back, Bruce and her parents saw that Alyson clearly had already shown herself to be particularly vulnerable to group pressures. When Bruce saw her after her parents bailed her out, she was contrite and desolate. She couldn't understand why she'd done it. She'd gotten in with the gang girls gradually, after she'd been dumped by her Goth boyfriend. At her school, the marginal teens— Goths, biker kids, wannabe gangsters, druggies—hung out in the same part of the cafeteria, and some were in classes with her. There was some overlap between the biker girls and the gangster girls—the boys they knew often worked repairing cars and motorcycles. No longer feeling comfortable with the Goths, Alyson found herself moving from the biker chicks to the gang girls. Although her friends weren't actually gang members themselves, they had sisters and cousins who were the real deal—most of whom, unsurprisingly, had dropped out of high school. Alyson's friends idolized and copied their big sisters.

On the day of the fight, Alyson was terrified. She and five of her friends waited outside a movie theater in a mall parking lot for Lina, their intended victim, to appear. Apparently, Lina had somehow crossed the members of the real gang—and they wanted to send a message that her actions would not be tolerated. Alyson's group cornered Lina and soon she was on the ground, crying, as they kicked and punched and slammed her back to the ground when she tried to get up. All Alyson could recall at first was her own heart pounding, the rush of adrenaline and fear. She didn't think, she just acted.

"I mostly yelled a lot," she told Bruce, trying to explain that she didn't really mean to hurt Lina. Alyson did hit the victim, she eventually admitted, but said that she'd tried to pull her punches and didn't kick as hard as she could have. Nonetheless, Lina had been hospitalized with serious injuries, including black eyes and a punctured lung.

As Alyson later learned, the activities she took part in didn't mean she'd even been admitted into the gang. The younger girls were really errand runners, not privy to the gang's real fights and initiations. Their older siblings and cousins wouldn't let the "kids" become involved in anything they considered serious. The younger ones copied what they could—and often, that meant violent behavior.

Afterward, Alyson was overcome with distress about what she'd done. At the time, she'd felt virtually nothing for her victim. In fact, she said, she didn't even see Lina as a person. On reflection, that made no sense to her. She couldn't really remember any thoughts at all, she said, still having difficulty believing that she'd taken part in a fight that hurt someone so badly. She'd always seen herself as a kind and caring person—how could she have done something like that? Lina had never done anything to hurt her. Bruce tried to explain some of the research about the power of groups to her, hoping to help Alyson understand and face up to her behavior.

Asch and Milgram's social conformity experiments suggest part

of the explanation. Another part is that in adolescents, the brain regions involved in empathy are undergoing remodeling and growth:
teens, particularly in groups, are often less empathetic than adults; it
takes them longer to consider the effects of their actions on others,
and the frontal areas involved in empathy and planning in adults are
less active when teens make moral choices.[15] Understanding how empathy depends on group membership offers further insight. One
marker of group membership is often race. A recent study compared
the reactions of white and Chinese people to images of people of both
races.[16] In one picture, a white woman was shown in a painful situation, her cheek pierced by a hypodermic needle; the alternate showed
a white woman being touched on the same cheek with a Q-tip. Virtually identical pictures of a Chinese woman in these situations were
also shown. The subjects were asked to rate how much pain they
thought the woman might be experiencing in each situation.

Their verbal responses were identical: both white and Chinese
people gave the needle picture the same pain rating, regardless of the
woman's race. But when Chinese subjects saw the Chinese woman
being pierced, their brain's anterior cingulate cortex was much more
strongly activated than when they saw the white woman in pain. The
same was true for white people who saw the white woman suffering.
This region was not activated at all in the Q-tip pictures—nor did any
subjects rate that as painful. The anterior cingulate is important for
perceiving the emotional aspects of pain—and is known to be activated when people empathize with others in pain. So, although people
consciously recognize the pain of people of other races, this research
suggests that brain regions involved in less conscious emotional reactions are more highly tuned to pain among people in their own group.

Although this finding has dark implications, other research suggests that our unconscious responses to racial information are more
malleable than you might expect. For example, when people are seen

as "being on the same team," there is evidence that race fades in importance, even on unconscious measures. One study involved people viewing videos of two teams of four men, each group including two whites and two African Americans. The men were talking about a conflict between the two teams—and their support for their own side. In one video, the men who were allied with one another wore the same color shirts; in the other, there were no visual cues as to which side someone was on.

After viewing the videos, subjects were asked to look at pictures of the men and connect what each man said to his picture. Unsurprisingly, other studies find that when the brain filters sensory information for relevance, it is more likely to make mistakes about elements it has dismissed as being less important, compared with those that appear critical. When there were no visual cues about allegiance, subjects paid significant attention to race and they were twice as likely to make mistakes about which side a man was on as they were to misidentify his race. But when men on the same team dressed alike, race became much less important and errors about allegiance fell.[17] Other studies have shown that simple familiarity—just being with and knowing people from other races well—also reduces unconscious racial biases. The more familiar we are with different types of people, the less likely they are to spur the unconscious stress response that our brains produce automatically when some physical characteristic doesn't match those we've categorized as known and safe.

Group loyalty is thus more flexible and fluid than it might initially seem. Another study that supports this idea looked at fans of rival English football (soccer) teams, Manchester United and Liverpool. Team and city loyalty is taken seriously there. Liverpudlians and Mancunians have historical resentments against each other, and injuries and even deaths have resulted from football-related conflicts. Not surprisingly then, after a discussion about the merits of their own team,

Manchester fans were more likely to later help an injured person wearing a Manchester United T-shirt than they were to help someone wearing Liverpool's shirt or an unbranded shirt. But here's the interesting part: if the fans were first told to consider their overall love for football as a sport, they were equally likely to help fans of either team—but not those wearing the plain shirts.[18]

Alyson's own case was a good example of how race can fall in importance in the face of familiarity and group solidarity, at least to some degree. She had been unusual as a white member of a Latina gang clique, but she had been fully accepted, at least by the "junior members." They would never have enlisted her for their attack on Lina otherwise. However, when the younger girls had taken her to meet their hard-core sisters and cousins, they'd made clear that as a white girl, she wasn't welcome.

WITHIN A FEW months, Alyson couldn't imagine what had possessed her to befriend the gang girls and why she had participated in the assault. As a Goth, she had demonstrated against war; unlike her former friends, she did well in school and her goals in life were completely different from theirs. She didn't even really like their music or style. But she needed to fit in, and she found herself among them when she no longer felt safe or accepted by the Goths. So, as she'd done before, she changed her colors to blend in. She felt she needed to belong somewhere. Being an outsider was much worse than being part of a "bad" crowd.

And that's the perspective of almost all teens. Fitting in matters more than anything else. As a result, although schools and governments repeatedly design and utilize programs to try to help adolescents "resist peer pressure" to engage in risky or violent behavior like Alyson's, very few such efforts are highly effective. Teen drinking and teen drug use remain with us, as do all kinds of other activities that parents would love to prevent. Unfortunately, many programs imagine peer pressure as an overt, external force, having teens engage in role plays where they "just say no." In fact, most peer pressure is subtle, internal, and sometimes unconscious. Many youths "say yes" without ever having been asked a question or pressured at all: they simply mirror the behavior of others. At this age, in these contexts, friends matter more.

In fact, research that has followed teenagers into adulthood finds that those who follow the crowd tend to be healthier than those who ignore it. Adolescents who care about how other people see them have more friends, better relationships with their families, better grades—and more empathy.[19] Those who don't worry much about their peers' opinions are the ones who are likely to pull *others* into misbehavior. Indeed, being unconcerned about those around you is basically the definition of being antisocial. Such teens don't care that their parents don't want them to drink, take drugs, or have sex. If their friends think they are "crazy," it doesn't bother them. These teens tend to find and egg one another on. Problems occur—as in Alyson's situation—when adolescents who do not have histories of antisocial behavior fall in with crowds of teens who do and copy their behavior.

Programs to prevent risky teen behaviors that work to any extent at all recognize this. Rather than trying to fight peer influence, they attempt to use it positively. For example, instructors emphasize the fact that teens who engage in binge drinking are a minority, rather than trying to get teens to fight against the crowd.[20] Instead of grouping troubled teens together, effective programs work to keep them in the community and place them in situations where youths who are doing well can serve as role models.[21] They keep the antisocial youths from making one another worse.

Because ultimately, constantly resisting group pressures—especially for teens—is like resisting sleep. Sooner or later, no matter how hard you try to continue to stay awake when tired, you will nod off. We need our groups to sustain us, but if you put yourself in a group at odds with your own values, your personal standards and beliefs are likely to be eroded by group pressure. This doesn't happen immediately, and people are rarely aware of the small and insidious steps involved, but over time, lines get crossed and before they realize it, they have drifted into patterns of behavior that their earlier selves might find unrecognizable.

In fact, one of the hottest areas of research right now involves studying social networks to understand how they influence behavior, in adults, not just teens. For example, a recent analysis of the Framingham Heart Study—which has followed fifteen thousand people for more than fifty years, trying to discern the causes of heart disease— found that behavior is, basically, contagious. The study showed that if one of your friends becomes fat, you have a 57 percent greater chance of gaining weight yourself. Oddly, even having a friend of a friend put on pounds had an effect: your risk of obesity rose 20 percent, even if the friend in the middle didn't gain weight at all.[22] Smoking, quitting smoking, drinking—even happiness and loneliness—showed similar patterns.[23] Through our propensity to mirror one another, even people we don't know can affect our behavior because they affect what our friends do and how they see the world.

And so, like Alyson, even the kindest, most sensitive person may behave ruthlessly under the influence of peers: again and again, history has shown that in-group pressures can erase empathy for outsiders. In fact, greater empathy here may sometimes be a disadvantage; it can make highly empathetic people more vulnerable to peer influence because they are better at picking up mirroring signals. Alyson's experience was far from an exception. Choosing one's friends wisely in the first place is much easier than being a lone voice of reason in the crowd.

This is not to say that no one can ever resist, of course. A study of non-Jewish Europeans who risked their lives to rescue Jews under Nazi occupation during the Holocaust found several key differences between this half a percent of the population and those who lived in the same regions but did not help. Those who helped had closer families, with parents who emphasized caring and altruism. They had befriended Jewish people at work or elsewhere—while nonrescuers kept their social circles closed and homogeneous. Rescuers also felt more kinship with a broader community that encompassed multiple social classes and religions.[24] These factors all help reduce "us-versus-them" thinking and failure to empathize.

Nonetheless, going against the crowd is difficult, and clashing repeatedly on matters important to a group's identity is near impossible if you want to sustain the relationships. Alyson knew she couldn't continue to hang out with the gang girls if she wasn't willing to engage in violence. She went through one of the most difficult periods in her life after the arrest. She became a complete outsider, a member of no clique or group. She felt utterly friendless. Slowly, mostly in solitude, but with Bruce's assistance and support from her family, she began to create her own identity and figure out what she genuinely enjoyed. He knew they were making progress when she joined the drama club and began to be cast in increasingly major roles in her school's plays. Onstage, she could put her mimicry skills to good use. In the drama club, she also found friends who—while they dressed wild and appreciated her intensity—were actually committed to school and their dreams of acting.

Alyson was extremely lucky, however, in that her parents were able to protect her from the worst consequences of her role in the assault. She had to do community service—which she threw herself into gladly—but aside from the hours immediately after her arrest, she did not spend time in juvenile detention. Unfortunately, juvenile prisons and other settings that concentrate "troubled teens" often serve to solidify and enhance teens' identities as gang members—rather than allowing them to move on. One recent study followed over seven hundred at-risk teens in Montreal, some of whom were sent to juvenile prison. When compared with teens with similar criminal histories who were not sent, those who were incarcerated were seven times more likely to be arrested as adults.[25] In Bruce's practice, another female patient who was just starting to become involved with a gang was sent to juvenile detention after having participated in a similar assault. She emerged with gang tattoos and later participated in a gang-related murder.

WE IGNORE OUR social nature—and our propensity to form groups, both positive and negative—at everyone's peril. Although empathy for "people like us" may not have originated in acts of hatred and warfare against "people like them," it is deeply influenced by group identity. Not just teens but also adults need to stay aware of group influence on values and choose friends, other social groups—even employers— with care.

nine | US VERSUS THEM

WHEN BRUCE SAW THE LITTLE BOY in his office neatly filling in a Disney character in a coloring book, he thought he should look at the chart again, just to be sure he had the right file. Surely this wasn't a twelve-year-old? The thin, wiry boy was wearing a perfectly pressed shirt, buttoned all the way up, with similarly wrinkle-free blue dress pants. He looked like he was ready for Sunday school—or waiting to have his school picture taken. He sat at the little table that was set up for preadolescents and spoke politely in a quiet voice. He seemed—mentally, physically, and in his behavior—to be more like an eight-year-old. But if the file was correct, this was Terrell,* the "predator" whom the editors of the newspapers thought should be tried as an adult. He'd allegedly shot another boy from his neighborhood—over a pair of Nikes. Then, he'd collected the shoes from the victim's body and had been wearing them when he was arrested.

Bruce's job was to evaluate Terrell, to try to explain why this seemingly ordinary little boy had committed such a heartless crime. Bruce was at the University of Chicago at the time, in the late 1980s. A rash of "senseless" shootings over jackets, sneakers, and "dirty looks" had hit the south side of the city, in the wake of the crack epidemic and the loss of manufacturing jobs and high unemployment that preceded it.

In Terrell's neighborhood, the odds that a young black male like him would be killed were 7 percent per year—many times greater than those for a U.S. soldier now serving in Iraq.[1]

Bruce had been approached by Terrell's defense team, but his evaluation would likely be available to the prosecution or others with legal standing in the case as well. What he found was a child who had been in and out of the child welfare system and, unlike Alyson, with all her middle-class advantages, had essentially been born into a gang. Terrell's story shows another possible path that can be taken when children's needs for empathy and connection are not met. It also illustrates what gangs can teach us about what people need from groups and the more subtle ways in which they can shape our behavior.

Terrell's mother, Jenetta,* was deeply troubled. A teenaged, single mom who was herself the child of a teenaged single mother, she lived in a world where it wasn't possible to be "neutral." You had to have people—and where she lived, the people who dominated were the Bloods. Jenetta was estranged from her biological family; she'd grown up in and out of foster care. Her mother was an alcoholic who eventually became homeless, and she never knew her father. But the Bloods were there, and while Jenetta wasn't good at sustaining romantic relationships, she stayed true to the Bloods. Stepfathers came and went in Terrell's life, but they were always tied to the gang. His biological father had been a member, as was his father. The group had a thriving crack trade in Chicago.

When Terrell was a baby, Jenetta began smoking crack. This pushed her already limited parenting skills to the breaking point. Consequently, Terrell didn't get the responsive, nurturing attention that babies require to smoothly regulate their stress response systems. Jenetta only infrequently mirrored his smiles and coos. When she talked to him, it was mostly to say no.

Both lack of nurturing and limited exposure to language can have a serious impact on brain development. Nurturing interactions lay the foundation for language development and for key aspects of self-regulation, including attention, aggression, and impulsivity. But language exposure matters, too. Verbal ability appears to be strongly related to impulse control; we encourage toddlers to "use your words" when they get frustrated for a reason. Language builds cortex—and the cortex modulates the lower, more reactive brain regions.

Unfortunately, research has found that children from poor families, like Terrell—even if the parents aren't single or struggling with addictions—hear far fewer words every day than those raised by middle-class professionals. Researchers Betty Hart and Todd Risley painstakingly studied the fine details of verbal interactions in parents' homes. If you think of a conversation as a dialogue, an "utterance" is one side of that dialogue—anything from a barked order to a lengthy digression. On average, every day poor children have about 178 "utterances" directed toward them per hour—while children of professionals hear 487 utterances in that time span.

To make matters worse, Hart and Risley found that most of the words directed toward the poor children are discouraging—"no" and "stop it" rather than "Good boy" or "Good girl." In poor homes, for every word of praise, researchers found three admonishments. Meanwhile, in the professional households, the children got six "great work!"s for each "don't do that!" By the time they start preschool, children in poor homes have half the vocabulary of children raised in professional homes—525 words versus 1,100—as well as increased vulnerability to stress and, possibly, reduced impulse control.[2] About a third of the differences between people on test scores for verbal ability can be accounted for by socioeconomic status.[3]

The situation is even bleaker for children like Terrell, whose sole parent is addicted to drugs. Jenetta's days were spent chasing the next hit; if she crawled on the floor with Terrell, it was to search the carpet for missing crack, not to play. Her own high stress levels from soaring and crashing, from constantly craving and worrying about money, gave her little chance of calming her own brain, let alone guiding Terrell's developing nervous system and vocabulary.

And as Jenetta's crack use increased, she paid less and less attention to Terrell. He was left alone for hours, underfed, his diapers infrequently changed. Eventually a neighbor called Child Protective Services (CPS), and Terrell began the first of several trips into foster care when he was five or six. In his records, Bruce could see that there were more than half a dozen reports of abuse or neglect filed in his case—all of them confirmed and labeled "founded."

In foster care, Terrell was placed with families who had many children, either biological, foster, adoptive, or some mix. There, too, he didn't get much individualized attention or instruction. Because he was quiet, he didn't stand out. Upset by the loss of her son, Jenetta went into rehab, and within a year, Terrell was back home. But Jenetta's attempts at recovery, unfortunately, usually didn't last long. She'd make it for nine months or so—typically under some kind of legal or child custody pressure to get treatment or stay in treatment. After that supervision ended, she would drift back into using. When she declined to the point where she was neglecting Terrell, another CPS report would be filed and he'd be back in foster care. He never developed a sense of stability or connection to any of his foster parents; whenever he got attached, he'd soon move for some reason or another. And when he was home with Jenetta, her ability to be there for him didn't last very long, so his connection to her was tenuous as well. Like Ryan, he never learned that people could be consistently caring.

The only stable presence in Jenetta and Terrell's life was the gang— their crack business, the men who belonged, and Terrell's friends in the neighborhood who were on their way to belonging. In school, Terrell was soon identified as a troubled, at-risk boy because he cut classes frequently and was never able to succeed academically. Having started at a disadvantage, he kept falling further and further behind.

He wasn't seen as a violent or aggressive, though, nor was he disruptive or a troublemaker. He kept his head down and tried not to draw attention to himself, positive or negative. He'd learned in the foster homes how to fly under the radar and keep adults from getting on his case by saying "please," "thank you," and "sorry." But in the special-ed programs he was soon shunted into, not being much of a problem also meant not getting much of an education. The teachers and aides had to deal with the more obvious chaos caused by the defiant children. This segregation from the mainstream also meant that Terrell was placed among many other youths like him, most on edge and underparented, many already junior gangbangers.

"So what happened with John?" Bruce asked, prompting Terrell to describe his relationship with his victim after they'd spoken for a while about school and other relatively neutral subjects.

"I capped him," Terrell said, without remorse, pride, or any emotion at all. It was as though he was talking about doing—or not doing— his homework. Through further questioning, Bruce learned that John—who was about twice Terrell's size—had somehow insulted Terrell and that Terrell had decided to take his sneakers. Terrell never went anywhere without his gun and he knew that if he showed it, he had to use it or he'd be seen as a punk. There was apparently no gang-related "beef," just two angry teenagers whose conflict escalated before either could consider the consequences.

Terrell was unprepared to respond in any other way. His childhood of neglect and chaos had made him hypervigilant to threat—he could perceive even the tiniest hint of aggression in someone's face, voice, or manner. Everyone's brain is wired with some degree of bias toward falsely detecting danger rather than avoiding the opposite mistake.

For Terrell, however, this effect was enhanced because in his world, failure to react quickly to threat was indeed potentially fatal—as it turned out to be for his victim. Studies show that people exposed to extreme threat regularly develop faster reaction times, and this was almost certainly true for Terrell. He'd seen what happened to people who didn't react quickly and powerfully.

To make matters worse, under severe stress everyone becomes less rational, less empathetic, and more impulsive. As we've seen, in these situations, the brain reduces activity in the frontal cortex to allow the faster, lower regions of the brain to take charge. Unfortunately, this means that decisions made under significant stress are not likely to be informed by careful evaluation of possible long-term outcomes— such as "If I kill this guy, I'm going to go to prison and mess up my life." Nor are these actions likely to involve empathetic considerations such as "If I were him, I wouldn't want to get shot." All that matters in that moment is evading or defeating the source of the threat.[4]

Extreme stress changes the way we respond to the world—highly threatening situations make it physiologically much harder to think clearly and to be kind. The more distressed someone becomes, the less able the higher, more considerate regions are to be active and take control. Also, because, like muscles, the brain regions that are used most become the most developed, chronic stress makes the brain more likely to dwell in states of high reactivity and low contemplation. So for someone like Terrell, frequently experiencing a state of fear can train the brain to react more impulsively. For all of us, our level of arousal—from calmness to slight apprehension to fear and then all-out panic—affects our ability to make good choices. (See Table 1 in the appendix.)

Understanding this "arousal continuum" reveals a lot about the way the world works—both in situations like Terrell's and in less extreme contexts. In contrast to threatening situations, safety allows consideration—for others and of a given action's future consequences.

If we want a kinder, more caring society, people need more experiences and places in which they feel safe. If we want to be kind to others or have others respond with empathy toward us, we need to minimize unpredictable and highly tense situations and maximize our ability to deal with ordinary stress. To encourage compassionate action, we need to create conditions and emotional states conducive to it. And to do this in the long term, we need to nurture mothers so that they can help their babies effectively develop the capacity to modulate stress. This can help end the cycle in which threat induces fear, aggression, and more threat and begin a healthier pattern where empathy produces kindness, and calm creates safety, which then allows more empathy. (See Table 2 in the appendix.)

For Terrell, sadly, there were also several other important factors driving his impulsivity and apparent failure to consider either his own or his victim's future. Importantly, at twelve, *no one* has a fully developed frontal cortex. The wiring of the brain's decision-making region takes place during adolescence and is not complete until the early twenties for most people. Not coincidentally, the likelihood that someone will commit a violent crime peaks during these youthful years—and falls at exactly the same time as cortical development is normally completed. Moreover, not only was Terrell's brain unlikely to have developed adult competence at impulse control by age twelve, but his own growth and development was delayed because of the neglect he had experienced. He was not only the size of a typical eight-year-old—but also had the same level of self-control. His stature reflected his developmental delay. He didn't sit down and color with Bruce to try to look innocent—he did that naturally because he was mentally at an age at which children find it appealing and appropriate. Unfortunately, that made his access to a gun even more dangerous.

And so when John said "fuck you" instead of giving up his sneakers, Terrell didn't hesitate to shoot him in the chest. In his world, this

wasn't senseless, it was necessary. Ta-Nahesi Coates describes the
role of "saving face" in a similarly tough neighborhood in Baltimore:

> Painfully, I'd come to know that face must be held against
> everything, that flagrant dishonor follows you. . . . Nowadays,
> I cut on the tube and see dumbfounded looks when over some
> minor violation of name and respect, a black boy is found
> leaking on the street. The anchors shake their heads. The
> activists give their stupid speeches, praising mythical days
> when all disputes were handled down at Ray's Gym.
> Politicians step up to the mic and claim the young have gone
> mad, their brains infected and turned superpredator. Fuck
> you all who've ever spoken so foolishly, who've opened your
> mouths like we don't know what this is. We have read the
> books you own, the scorecards you keep—done the math and
> emerged prophetic. We know how we will die—with cousins
> in double murder suicides, in wars that are mere theory to
> you, convalescing in hospitals, slowly choked out by angina
> and cholesterol. We are the walking lowest rung and all that
> stands between us and beast, between us and the local zoo, is
> respect, the respect you take as natural as sugar and shit. We
> know what we are, that we walk like we are not long for this
> world, that this world has never longed for us.[5]

Terrell moved in a world where he had to act—or be acted on. Unlike Alyson, he hadn't acted under the immediate sway of a group. Instead, he had internalized his group's rules, which he described to
Bruce: stay loyal, accept no disrespect, and we'll take care of you. This
is our creed: you don't ever rat, you don't ever steal from us. As for the
rest of society's morals and laws, they don't matter. The law doesn't
protect us, so we don't respect it. And we should have good things, as

much as anybody else. Why shouldn't I have that coat? Those shoes?
I'm going to take them.

Terrell knew his victim played by the same rules; he knew he
might die in the same way if bad luck or poor judgment about an adver-
sary put him on the wrong end of a gun. He was fatalistic. It was just
how it was. He'd seen people stabbed, seen people shot. If it was your
time, it was your time. Unlike Alyson, he'd never known anything else.
That is not to say he didn't bear any responsibility for his actions—but
his choices were shaped and constrained by his upbringing, his neighborhood, his poverty, and his seriously delayed and immature stage of
brain development. In fact, imaging research showing the brain's im-
maturity during adolescence recently led the Supreme Court to ban
the death penalty for juveniles.

Bruce asked Terrell about where he lived and his family background. Terrell responded with a long list of nicknames and first
names. When Bruce inquired further, he discovered that the people
he named weren't blood relatives, just Bloods. It turned out that Terrell had stopped living at home a few years earlier. His mother had
begun turning tricks; her pimp didn't want him hanging around, a boy
his age brought too much potential for trouble. So, while he wasn't visibly homeless, he didn't have a stable home, either. He'd spend a few
weeks with one friend on his couch; another few in the spare bedroom
of another. His mother might call from a new apartment and then he'd
be back there for a few months—until a new man came along or his
mother's addiction deteriorated and she stopped paying the rent.
Then, it was back to the ad hoc housing program the Bloods provided
for their up-and-coming soldiers and corner boys.

"What do you think is going to happen to you?" Bruce asked.

"I'll be in juvie," Terrell said, seeming completely unconcerned. "I
have people there." At first, Bruce thought he was in some kind of de-
nial, completely disconnected from the consequences of his actions.

But as he learned more, he realized that to Terrell, juvenile prison might look like—might actually be—a step up, or at the very least, not worse than his current situation. In juvie, he'd have his own space and not have to constantly move from house to house, relying on others for shelter and sometimes food, negotiating treacherous, ambiguous debts with people he'd eventually have to pay back somehow. From his perspective, he knew that all he had to do in juvie was show the right gang signs when he was walking down the hall to lunch, and there they'd be, the guys who grew up with his father and his cousins—his people.

Unlike Alyson, Terrell didn't have a supportive family or any other place he could fit in. Unlike Trinity, he didn't have the advantage of high sensitivity, intelligence, and at least some caring neighbors or teachers who had been able to reach him. He'd adapted to a cruel world that cared little for him, or for his victim. His lack of empathy echoed the lack of care with which he'd been treated. And so, his gang became his family. Like a family, it provided shelter, companionship, and an established set of values. Although this put him at odds with both the law and other gangs, it was one of the few visible ways boys like him could get any of their social needs met, if minimally. What may be most shocking about his story is not the way he behaved, but perhaps that violence isn't more common in neighborhoods where there are so few alternatives for boys in his situation.

TERRELL DIDN'T JUST "end up" as a Blood—he slid down a steep path into their arms that had been taken by his father and his grandfather before him. His story reflected a transgenerational pattern related to the loss of vibrant cultural and family bonds in inner-city Chicago. His family hadn't always been fragmented and dysfunctional—that had only affected the most recent generations. But all of us are influenced

by patterns of cultural transmission that change as the groups we belong to connect with, colonize, or are colonized by others.

The human brain has a remarkable capacity to learn, discover,
create, and invent—and then, most remarkably, to efficiently pass these
discoveries and inventions on to the next generation. Our brains allow
us to absorb and to challenge the accumulated wisdom of thousands of
previous generations, enabling our species to repeatedly reinvent how
we live together, work, raise our children, and govern ourselves. Sadly,
when cultures meet, exploitation and destruction have been at least as
common as mutual benefit.

These days, life in a typical Western household is dramatically different from life in an indigenous hunter-gatherer clan. Each human
living group, however—each family and each culture—has some set
of unifying elements that help keep the group bonded. These include
language, beliefs and values, child-rearing practices, and other rituals
and traditions. The primal relationships between parents and children
develop in the crucible of these cultural practices—and the health and
welfare of the group around them depends on how well they support
families.

Consequently, breakdown of key cultural structures can rapidly
affect the number, quality, and nature of our relationships. To make
matters worse, the loss of these social anchors typically results from
other highly stressful events such as war, famine, or economic depression. The distress that this causes to individuals—combined with the
decrease in the buffering capacity of healthy relationships—can have
a toxic impact on the creativity, productivity, and health of the dissolv-
ing community.

This can be seen most clearly in the histories of formerly enslaved
or indigenous peoples. Multiple academic disciplines have docu-
mented numerous emotional, social, and physical health problems
that increase when social fabric deteriorates. The history of Western

societies' interactions with these groups is characterized by policies of active cultural destruction followed by incomplete assimilation and disenfranchisement. The consequences are sobering. Suicide, violence, alcoholism, child abuse, academic failure, diabetes—rates of essentially any emotional, social, or physical health problem affected by the physiology of the stress response systems are higher in indigenous populations throughout the world. But a transgenerational process of healing can also take place in these groups as they reconnect to their cultural anchors and incorporate healthy elements of the modern world. This process is both hopeful and instructive.

"THIS IS THE Friendship Dance. Everyone can do it. It is just like Follow-the-Leader. Get a partner and just take your steps like you have a sore leg." The smiling, dark-eyed man limp-stepped around the drum circle to demonstrate. Ray "Co-Co" Stevenson, an Ojibwe musician, was leading this celebration. For twenty years, Ray had studied, practiced, and performed the traditional songs of his people. His wisdom, wit, drumming, and remarkable voice make him in great demand for powwows and other ceremonial events throughout Canada and internationally. Ray is highly respected and skilled.

Grandmothers, children, middle-aged men laughed at his silly steps as they stood to join the groups of Ojibwe, Dakota, and Cree dancers wearing full traditional regalia. Dancers had come from all over Manitoba to participate. All ages were represented—children, teens, men and women, young and old together. Varied and beautiful, the garb included beaded moccasins, dresses and shirts, full-feathered headdresses, anklets with bells, and ceremonial fans.

This powwow, a communal gathering of dance, song, and feasting, was part of a celebration of the twenty-fifth anniversary of a remarkable grassroots organization in Winnipeg, Manitoba. The Ma Mawi

Wi Chi Itata Centre (usually called Ma Mawi, pronounced "mom away") was started by members of the aboriginal community in Winnipeg to address the many problems related to the fragmentation of family, community, and culture seen in the urban aboriginal population. The powwow was officially started by a grand entry procession. The elders and dignitaries followed a set of honor guards in full uniform or traditional regalia.

Bruce had been asked to speak at the celebratory gala the night before and was invited to attend the powwow and join the elders and dignitaries in the opening processional. For one poignant moment as this procession began its respectful march around the large drumming circle, a skeletally thin Native man wearing filthy jeans, a ripped T-shirt, and a crumpled red baseball hat jumped into the parade. He was clearly drunk. He was gently led out of the circle by one of the elders. There was no shaming—everyone knew people like him in the community; some of the now-proud participants had themselves recovered from alcoholism and they knew that a relapse could easily put them in that position again. But the sad contrast made the power of the unifying and often dazzling dancing to follow even more obvious.

Today, the Ma Mawi had a lot to be proud of, however. Their programs had helped support and rebuild many families into a healthy community. The translation of their Ojibwe name literally states their core value: "we all work together to help one another." For the last twenty-five years, through a range of services and programs, they have changed many lives and helped reweave social fabric in the urban aboriginal communities in Winnipeg. Ray is a perfect example. In fact, at one time, he could easily have been the drunk who disrupted the ceremonies.

As he states, "I was all about sex, drugs, and rock and roll. I was on a bad road. So many of my friends are dead or in jail." He had suffered from family fragmentation, had lost loved ones to traumatic death, and had grown up surrounded by addictions. His early youth had been a predictable reflection of these developmental stressors. He'd always had problems in school. Soon, he connected with other marginalized, traumatized youth. Together, they'd discovered smoking, drinking, and drugs—and then they became involved with other crimes. Ray had actually been introduced to traditional singing by being adjudicated to one of Ma Mawi's programs.

Sitting in a traditional sweat lodge and hearing the singing and drumming around him, he was moved. The drum spoke to him. He knew suddenly that this was what he wanted, what he was meant to do with his life. He connected with several older, traditional singers. Ma Mawi had introduced "powwow clubs," which instructed these youths in Ojibwe song, dress, and dance. Ray was mentored. He had a gift that was recognized, nurtured, and guided by others. He was pulled into a healthy group, which connected him to his cultural past.

More than that, the singing and drumming had patterned, rhythmic qualities that could soothe his internal anxiety. The loss, chaos, and experience of threat in his youth had dysregulated his stress response system—and before learning to sing, dance, or play this music with his people, his only source of relief had been alcohol and other drugs. Now the music could reach the same parts of the brain that the chemicals did. Especially with patients who have suffered early maltreatment that affected nonverbal and unconscious brain regions like the stress response system, Bruce has repeatedly found that activities that combine music and movement can be powerfully restorative.

Ta-Nahesi Coates writes of this, too, as a healing activity during his teens in Baltimore:

> In his hands, the drum sounded like a gun, if guns were
> made to be music. The boy, only slightly older than me,

affixed it between his legs with the aid of a long strap, and
ever so casually began to make it sing. We were learning the
dance steps from the Mandé, the traditional gyrations made
to heal the insane, celebrate the harvest, or inaugurate a
tournament of wrestlers. I could not move. True enough, the
initial cause was great fear—everyone knew I danced as
awkwardly as I moved through life. But more so, I was held
by how the brother played, and how unconsciously it all came
out. It was like he had no plans. I could catch the basic beat,
but what he brought out of it showed that he heard more. And
as I listened, I became bewitched. . . . My breathing
quickened whenever the drumming began. I would bob and
nod unconsciously. My hands would move involuntarily.[6]

WITH A BETTER way of relieving his tension, Ray was able to stop
drinking and using. Two key elements of health converged in his new
found skills. First, the powwow club connected him to a community,
providing a wealth of supportive relationships and meaningful interactions. Second, the music and dance themselves helped him better
regulate his stress response. Rhythm and rhythmic movement are
critical to the basics of life: almost all of our regulatory functions keep
some sort of a beat, from the heart to the breath to the menstrual cycle
in women. Consequently, music and dance, with their echoes of the
normal patterned, physical interaction between mother and child can
be healing. And soon, Ray would be able to bring this healing to others. His insights and wisdom grew with time, experience, and teaching. He had been given guidance and now he would guide others. He
started to set up and run more powwow clubs for Ma Mawi. He now
plays in the band Eagle and Hawk. But he's always chosen to put his
work with children first. He continues to work to rebuild relationships

and community, support recovery, and strive for the harmony that originally characterized the core values of his people.

After the Grand Processional, the elders and dignitaries, including Bruce, sat down. The dancing began. Bruce appreciated the beauty and the proud displays of the dancers and the varied drumming circles (groups of singers and drummers who practiced and performed together). In fact, he calmly enjoyed everything until he was pulled up to join the Friendship Dance. Then, he felt like a clumsy, oafish outsider. He was anxious. Within a few minutes, though, his distress faded. The echoing drumming, the warm smile of the tiny six- or seven-year-old girl holding his hand, and the patterned, synchronous dancing made him feel connected and part of the group. This truly was a friendship dance. Cree, Caucasian, African American, Ojibwe, Dakota, elderly, young, male, and female—all were connected in that moment by touch, smile, song, drum, and dance. Eighty beats per minute was the central tempo: that primal rhythm that unites everyone in the memory of the maternal heartbeat in the womb, recalling the safe, warm, nurturing steps of our first relational dance.

Thus, it's not surprising that all cultures and societies have developed some form of patterned, repetitive, and rhythmic "dance." Dance has always been used to bring people together—both for good and for ill. There are not just rain dances or celebratory powwows, but war dances as well. In fact, some have theorized that the reason that every known human society plays music together—and the key role of chanting, singing, and dancing in so many group activities and rituals—is a result of music's ability to further bond people to one another.

A series of recent experiments showed that walking, singing, or moving in synchrony all increased people's cooperativeness. In one of these, strangers were asked to either march in step or walk normally—afterward, those who synchronized their movements were more likely to predict that a partner would cooperate with them to make money, and, therefore, they behaved cooperatively in a financial game them-selves.

Those who sang together while moving in unison were also more cooperative—even at a risk of making less money for themselves. In contrast, those who were instructed to be "out of step" physically with the song they were supposed to sing together were less cooperative than people who synchronized.[7] Because there were no other differ-ences between the synchronized people and those who weren't, syn-chrony had to be responsible for the improved collaboration. The same neurology that allows empathy allows synchrony. This is one reason why military organizations still emphasize these exercises. Marching in unison, exercising intensely together, and chanting brings troops together and unites them as a unit. The complex choreography in some gang signs and body language is probably another reflection of this unifying quality of synchrony.

Of course, the flip side of empathy for "us" is hatred and dehuman-ization of "them"—without which crimes like Terrell's couldn't occur. Over the course of history, armies, gangs, and others with violent goals have learned many ways to heighten altruism among "us" while encouraging barbarism toward "them." One of the most common is the use of dehumanizing language: calling enemies "cockroaches" or labeling them as other disgusting animals or objects. Virtually all rac-ist and genocidal propaganda includes such images; gangs also use degrading language to target their enemies. The less the enemy is seen as human, the easier it is to avoid compassion for him.

In Terrell's case, however, his state of fear and his immature brain were probably the most important factors in enabling the killing. His experience of being "disrespected" had immediately made his victim a threat. And being threatened, poised to defend his group against disrespect, armed, and a child all combined to produce tragedy. As he predicted, Terrell was ultimately sentenced to juvenile prison: the calls to charge him as an adult faded as other crimes garnered media attention. So at least one thing had gone his way: research shows that children sentenced as adults are about 34 percent more likely to commit new crimes after being released than those sentenced as youths.[8] Placing youths together with hardened criminals only furthers their connections to crime—it doesn't scare them straight, and it makes them vulnerable to predatory behavior that can further traumatize them.

HUMANS ARE UNALTERABLY social beings. We need family, we need friends, and when children are raised in tough environments, these needs become even more important, not less. Terrell's story reminds us that when children's needs for care aren't met, they don't just go away. Whether with blood relative or Bloods, teenagers will find somewhere to belong. The values of the group they choose can become irrelevant to the child in the face of that need. But they are not irrelevant to the rest of us—they can set the tone for whole neighborhoods, even societies. It's not just gangsters who get caught in the cross fire.

ten | GLUED TO THE TUBE

I THINK YOU NEED TO BE a little more specific here," Bruce said to the resident, who had sent in a form seeking a psychiatry consultation and was now speaking with him on the phone. When he was chief of psychiatry at Texas Children's Hospital, a big part of Bruce's job was providing "consults" when other doctors had a problem they couldn't address. It worked like this: a child would do something strange or troubling, like throwing poop at people or randomly biting staff, and the patient's doc would call psychiatry. A secretary would fill out a form describing the situation. In this case, the sole "presenting issue" on the paper he now had in front of him was that the patient, a five-year-old boy named Brandon,* was a "pain in the ass."

After stammering a few stories that didn't make much sense, the young doctor said, "I can't put my finger on it, Dr. Perry. But I know that there's something wrong with this kid." He described how Brandon would appear to understand what he'd been told, but then would continue to do things like removing IVs or picking at wounds. He didn't seem to be defiant, but he didn't comply with doctor's orders, either.

And it was unusual: most children, if you tell them not to touch an IV because you will have to insert it again, will leave it alone. They don't like it when you poke them and they don't want to go through it a second time. But Brandon was different. He kept pulling it out, and he would get up and wander around and take things from other people's rooms. Something was up. Bruce arranged a time to meet with the resident at Brandon's bedside.

When he got there, he saw a skinny towheaded boy with big blue eyes. He was sitting and staring at the TV over his bed, slack-jawed with the glassy look so many children have when they are immersed in a show or a video game. Unless he opened his mouth, he looked ordinary; in fact, he was beautiful. Bruce knew that Brandon was in the hospital for extensive dental work, but that didn't quite prepare him for what he saw when he smiled. Brandon's teeth looked like those of a methamphetamine addict or aging junkie: they were so decayed and rotten that many of them would have to be pulled.

Bruce walked over and knelt down by his bed to talk to him. Brandon was so immersed in the cartoon he was watching that he didn't even look at Bruce. After a few moments of being ignored, Bruce looked at the boy and turned the TV off. He didn't bat an eye: Bruce had expected more of a response, possibly an angry one. But now Brandon sat and looked at him, blankly.

Bruce gave Brandon his usual explanation and introduction, telling him who he was and that he would be back in a few minutes to see if he wanted to talk. Bruce always begins this way, saying that the child doesn't have to talk and giving him a sense of control over the interaction. This allows children time to think and prepare to speak with a stranger; if you just walk in and start talking, they are taken by surprise and are, reasonably, focused on who you are and if they should trust you. They respond in a protective and defensive way, rather than in a manner that reflects their real capacities. Our brains are even more likely to perceive new people as threats in a scary setting like a hospital. Bruce always wants to talk to a child who is as calm as possible. And when given a choice to talk after they've had a chance to think, children almost never decline.

But Brandon's responses, even to this normal introduction, were unusual. He seemed chirpy, but weirdly robotic. "So pleased to meet you," he said, smiling, and Bruce was struck by the overly formal phrasing.

"How're you doing?" Bruce asked.

"New and improved," Brandon said, with the same too-bright tone. Bruce gave him the usual spiel, ending by saying "I'll be back in a few minutes." He later realized that this must've cued his next response, but at the time, it just baffled him.

"Supersize me!" Brandon exclaimed, then returned to a dull, glazed stare. It was as though a switch had been thrown to shut him off.

OK, Bruce thought, now really curious about this boy. Was this why the resident had said that Brandon was a pain in the ass, was this his way of being a smart aleck? *What is he trying to tell me?* Bruce went to look at his chart. Over time, by looking at his records and talking with Brandon's mother, he pieced together a devastating story of a child who had been raised by a TV—not just as a babysitter, but as a parent.

As an extreme case, Brandon's story can tell us a great deal about how TV and other screen time like video-gaming and Internet use affect the development of empathy—and how a mom's depression can have unexpected consequences for her child. When Bruce had said he'd be back soon, Brandon had heard "More after this!" and responded with his version of a commercial break.

LIKE MANY—PROBABLY most—parents who neglect their children, Brandon's mom did not intend to harm him. When Bruce spoke with her, she was terribly ashamed of the state Brandon was in and the way she'd failed to care for his teeth. Her depression was palpable: you could see it in her hunched posture, her lowered eyes, her monosyllabic replies to questions, and the exhaustion and effort that were audible in her voice as she tried to respond to Bruce. The depression emanated from her like a bad odor, filling the room with its heaviness and her weariness. Even spending a short time with her was painful for a reasonably healthy adult—it was frightening to think about what it was like to have her as your only contact with the world, as a helpless baby.

Lacy* was ignorant about the developmental needs of children, but the main problem was that even when she knew she wasn't doing enough for Brandon, she was simply incapable of doing more. She was twenty-six, a single mom who had no relationship at all with her child's father. Getting out of bed was a tremendous act of will for her each morning; keeping her child fed took all of her strength.

The lethargy extended to all areas of her life. Sometimes, she did better than others and could hold a job for a few months. Then, the black fog rolled in again and it took all she had just to buy groceries twice a week. She had no friends or family and only left her apartment when there was no other alternative. Like many people with depression, she smoked heavily: it turns out that nicotine has some antidepressant effects, and this may be why despite all the antismoking campaigns, most people with serious mental illness continue to smoke. Often, Lacy would sit in her kitchen and chain-smoke, propping baby Brandon in his high chair in front of the TV with a juice bottle to keep him quiet. This practice was what had destroyed his teeth.

She didn't ignore his cries—at least not most of the time—but she didn't pay loving attention to him, either. She didn't rock him and hold him and play little games while she fed him. She didn't smile at him and play patty-cake or peekaboo. She didn't talk or interact much at all. She was there physically—but not mentally and emotionally. As a result, Brandon learned most of his words and language from the only other source of stimulation that he had, the television. And since commercials are usually louder and flashier than the programs that surround them, advertising jingles became his most important introduction to language.

IF YOU LOOK at the situation without understanding the social needs of the brain, it wouldn't seem like it would be a problem to learn mainly from television. Lots of words are spoken, they are linked visually with actions and objects, and there's a great deal of repetition. Of course, this disregards the primary function of language, which is to communicate with other people.

As we've seen, the brain is a social organ and requires affectionate contact to develop normally. Ordinarily, the interaction between mother and child teaches the stress system how to soothe itself, providing a calm state for learning to take place. This preverbal interaction also sets the stage for language learning in other ways, as it attunes the child to the rhythms of conversation and sets up the mirror neurons to copy not just the parent's words and actions, but her emotional state as well.

Now consider Brandon's situation, as he sat in a high chair in front of the TV. He's hearing many words and seeing images—but his responses to them don't matter. Whatever he does, the TV goes on blaring programs and advertisements, never changing in response to his reactions. In a normal situation, a child smiles at someone and the person smiles back; if he asks for food, his caregiver provides it. The TV, however, just sits there impassively. From the child's point of view, it ignores him. Brandon's brain did learn to make connections between certain words and certain images, but this didn't really allow him to understand how language is used, how tone of voice relates to emotion, and, most important, how to use words to form social connections. You can't learn empathy from something that can't empathize— and without reciprocity and interaction during teaching, language learning is severely impaired.

Brandon's mother's depression—aside from causing his neglect and overexposure to TV in the first place—also hindered his ability to attach and connect. To be sure, maternal depression doesn't automatically cause bad parenting. In fact, most depressed mothers are excellent parents. But maternal depression does make being a good parent much harder. If unaddressed, it can have a profoundly detrimental impact on infant social development. And unfortunately, maternal depression is an extremely common condition: depression itself occurs twice as frequently in women as in men.

In fact, there's one form of depression—which Lacy had—that is associated with childbirth itself. Postpartum depression (PPD) affects between 7 and 13 percent of women after they give birth.[1] It is much more common in single moms, mothers whose partners and families are not supportive, and poor mothers, particularly when they lack not only material resources but also close family connections and friends. Lacy qualified in each of these ways.

Evolutionary theorists suspect that PPD may be an adaptation that developed to disconnect mothers from children who are born into circumstances they are unlikely to survive. Research showing that PPD is more common in resource-poor, socially neglected moms and in moms of sickly, premature babies supports this idea.[2] It sounds harsh, but in the circumstances in which early humans lived, moms were more likely to have children who survived to reproduce themselves if they did not become too attached and devote too many resources to babies who were unlikely to make it to adulthood.

Instead, if they saved their energy for existing older children or to

reproduce again when food and emotional support were more available, they could reduce their odds of losing all of their children—not
just the one baby—to the high childhood mortality rates that characterized most of our evolutionary history. By detaching, these mothers
could start to protect themselves from the pain of losing the baby, but
the detachment itself almost certainly made the baby even less likely
to survive. As we saw with Eugenia, when children do not receive
enough physical affection, growth hormone production is turned off
and they "fail to thrive." Like her, Brandon was extremely small for
his age.

Because it is so common, maternal depression has been relatively
well studied. Research finds that children whose mothers were clini-
cally depressed when they were infants have poorer language skills[3]
and are more prone to later behavioral problems, depression, addictions, and other mental illnesses. Brandon's delayed development and
language problems were likely due to this as well as to his overexposure to television.

Many of Brandon's current problems had their roots in mirror
neurons and the role of empathy in development. Like Lacy, depressed
moms can be less responsive to their children's emotions. Her failure
to share joy and respond to Brandon's smiles with smiles likely made
him first confused, then anxious and depressed. Children of depressed mothers tend to be more withdrawn: their mirror neurons are
copying depression and reproducing it in themselves. Babies evolved
to be very sensitive to their environments, particularly to the emotions
of their primary caregivers: when their mothers are depressed, they
read this as a signal of severe stress and respond accordingly.

So some of Brandon's emotional and behavioral problems started
with his mother's illness. As he mirrored Lacy's distress and distance,
Brandon's ability to experience more positive emotions became muted.
When your mom can't even comfort herself, continuing to seek comfort

or implicit instructions on self-comfort from her simply invites more pain.

The effects of maternal depression on babies' development have been studied in miniature in an elegant series of experiments called the "still-face" paradigm designed by developmental psychologist Ed Tronick. During the experiment, mothers are instructed to look at their babies, but not respond with any facial expression of emotion. Even when only done for a few minutes, this is surprisingly stressful for both mother and child.[4] Typically, the experiment is done with three-month-old babies after the mother has settled the child in an infant seat, played with him for a few minutes, then turned away and then back toward the baby.

At first, the baby looks at the mom and smiles. When she doesn't respond to this, he starts to get anxious, looking to the side and then raising his arms imploringly. As the mother continues to stare stonily, the baby's movements get increasingly disorganized and distressed. Eventually, the baby looks away and withdraws into himself, sucking on his fingers or toes for comfort and looking terrified. Some cry. Most of the mothers in the experiments report feeling horrified— both by the distress they see in their babies when they don't respond, even for such a short time, and by the power they see they have over their children's emotional lives

Unfortunately, Brandon probably had numerous experiences like this early in life, where his mother's depression prevented her from responding emotionally, even when she was physically present. Please note that this does not mean if you don't respond to your child instantly or within minutes of a cry or other expression of need, you will harm him for life—the still-face situation is artificial in that the mother looks at but deliberately ignores the child. Most times when you "ignore" your baby, you are out of the room or looking away and babies rapidly come to understand that it's not possible for you to pay attention then. It's hard to accidentally create a situation in which you look into a baby's eyes and intentionally disregard her obvious pain! However, depression can mean that you pick up on fewer cues from your baby or are less able to share in his happiness. This is one reason why it is crucial to seek help for postpartum depression: not only does mom suffer, but baby does, too.

Brandon had experienced Lacy's failure to care for him so many times that he'd essentially given up. It was no wonder Brandon couldn't connect to anything other than the TV. He'd stopped expecting people to react to him—the expectation became too painful. Instead, he lost himself in the programs and ads. To help him, Bruce was going to have to first help Lacy—and then ensure that both got support from a wider social network.

The first order of business was to get her depression treated. Unfortunately, the way our health system usually works is that we hand someone like Lacy a phone number to call and expect her to be able to not only follow through on that, but also to show up for necessary appointments and deal with any related insurance or Medicaid hassles that might complicate them. It's somewhat like expecting someone who needs a wheelchair to walk to the store to pick it up. If she could do things like that, her child wouldn't be in the hospital now! So Bruce's staff took over Lacy's treatment, prescribing her medication and getting her started in talk therapy.

Brandon was referred to speech therapy to help him communicate more naturally. And as Lacy's depression improved, so did Brandon's speech and behavior. Because they lived far from the hospital, Bruce had to refer their care to other professionals, but the last he heard from them, both mother and son were doing well. Brandon's dental work was completed, and freed from frequent trips to the dentist and hospital, he began acting and speaking in a much more natural and less stilted way. He will probably always be distant and somewhat awkward with people and is at high risk for depression and behavioral problems—but because Bruce and his team were able to intervene with both mother and child, his chances for full recovery are much better.

SO WHAT EFFECT did all that television have on Brandon's brain? We now know that for the youngest children, screen time isn't associated with any positive effects. One study found that even normal infants exposed to supposedly "educational" baby videos had measurable language delays: each hour spent watching *Baby Einstein* or similar videos was linked with six to eight fewer vocabulary words compared to infants who didn't watch. Though *Baby Einstein* was named to sell parents on creating little geniuses, the effect of high exposure to these programs is more likely to be similar to Albert Einstein's actual childhood behavior: he didn't speak at all until he was four. (And indeed, Disney has offered a refund to parents who bought these videos, while not officially admitting any problems.)

Nonetheless, by three months, 40 percent of children are regular video and TV viewers; by two years, 90 percent watch two to three hours a day. On average, American children in general watch two to four hours of television daily; one study found that this meant that before graduating from elementary school, each child would witness eight thousand on-screen murders and more than a hundred thousand other acts of video violence.[5] Most children, of course, have parents who pay at least some attention to what they are watching and make some effort to limit TV. Brandon, however, watched eight to twelve hours a day or more. And his mom wasn't paying attention to what was on that screen.

Research on the effects of television watching and exposure to video violence is confounded by situations like Brandon's. Because parents who let their children watch TV without limit tend, like Lacy, to have other, serious problems, it's difficult to sort out which negative effects result from TV itself and which result from things like neglect, social isolation, poverty, stress, and maternal depression. However, we do have pretty good evidence that too much screen time—particularly when it involves violent content—can be harmful to children and reduce their capacity for empathy. Even though the interactivity of computer games and the Internet is different from the passive nature of watching TV, these activities still cannot provide the social engagement of face-to-face interaction and playing.

More research has been done on violent content on television and in films than on video games. For decades, studies have found a consistent link between viewing on-screen violence and aggressive behavior. This connection is strongest for the youngest children, and it is found in many different types of experiments. Since you obviously can't randomize children to either watch or not watch violent television for years, many have argued that these links are only correlations and that violent TV doesn't cause aggression. Instead, they claim that aggressive children simply prefer violent TV and would behave just as violently without that exposure.

However, that argument cannot account for the results of numerous studies in which children are assigned to watch either a violent or nonviolent video or film. In most of this research, those exposed to violence behave more aggressively immediately afterward. In one study, for example, 396 seven- to nine-year-old boys watched either a violent or nonviolent film and then played indoor hockey. Researchers who did not know which film the boys had seen rated their aggressive acts during the game, looking for moves that are banned in hockey. These included shoving another boy to the floor, elbowing, kneeing, and tripping opponents.

Overall, the boys who saw the violent film were more aggressive. And, in one condition in the study, referees carried walkie-talkies that had been seen in violent scenes in the movie. The highest increase in aggression was seen in boys who had previously been rated by their teachers as particularly aggressive and who saw both the violent film and the referees with walkie-talkies.[6] In other words, the children who were already most violence prone learned an association between a particular item and on-screen violence by seeing the film, and that association affected their real-life behavior. This is quite troubling.

Other research confirms that aggressive children are particularly attracted to violent media. Some argue that this is because it helps them justify their behavior; others claim that aggressive children tend to seek higher levels of excitement, novelty, and stimulation in general. Both are probably true, varying by individual cases.

Unfortunately, many studies—like the hockey research—show that children who are already aggressive are the ones who are most likely to be negatively affected by video violence. So those who say that the correlation between screen violence and real violence is due to violent children seeking screen violence, not screen violence provoking them, are correct, in part. Aggressive children are attracted to violent media—but violent media can also make them *more* aggressive.

Of course, there's a big difference between being more likely to elbow someone during a hockey game and committing real crimes— but studies that have followed youths over time have found that video violence is linked with real crime, too. For example, a study published in 2003, which followed 329 children for fifteen years, found that 11 percent of the men who had watched the greatest amount of violent TV in their junior high years had been convicted of a crime, whereas just 3 percent of the rest of the male sample had been. This group was also nearly twice as likely to have "pushed, shoved, or grabbed" their wife within the last year—and the women in the sample who watched the most violence were more than four times as likely to have "punched, beaten, or choked" an adult than the rest of the women. These correlations remained after controlling for parenting and socioeconomic status.[7]

An analysis that compared American cities in which television signals were available early in the 1940s to those that didn't get decent reception until the 1950s (and controlled for other variations between the cities by comparing children born before and after TV was introduced) found that for every additional year a child was exposed to TV before fifteen years of age, there was a 4 percent increase in the number of arrests for property crimes and a 2 percent increase in violent crime. The authors concluded that 50 percent of the increase in property crime and 25 percent of the increase in violent crime in the 1960s was linked to television.[8]

And it's not just physical aggression onscreen that affects children. Even some "educational" shows for the youngest children, like *Clifford the Big Red Dog,* aren't entirely innocent. In young children, in fact, these kinds of programs can increase aggression even more than those that contain obvious violence. How could this be? Perhaps because these shows avoid physical assaults and explosions, they set up emotional conflicts to provide drama. They turn out to demonstrate lots of aggression, but verbally, not through violence. Children who watch such shows are actually more likely to exclude or put down other youths than those who watch overt violence—and their physical aggression is increased almost as much as in those who watch programs that include it.[9]

Video-game research parallels the studies of earlier media so far. Both adults and children given the opportunity to behave aggressively after playing are more aggressive after playing violent games. And this aggression isn't simply excitement being expressed after a stimulating activity. When violent and nonviolent games that produce the same level of emotional arousal are compared directly, increased aggression is only linked to violent games.

Such effects are predictable given what we now know about the brain and the development of empathy. Children mirror the behavior of those around them—and this can include those they see in the media, whether the aggression is physical or relational. Violent video games may be particularly problematic in this sense for several reasons. Research has shown that people who identify themselves with the violent perpetrator in films and on TV are more likely to be aggressive.[10] In violent video games, this identification is more complete because you actually take aggressive action while playing that role. These games also involve repeated actions, which reinforces learning more effectively than passive, one-time-only behavior.

Of most concern here is desensitization, which is also called "tolerance." This is the numbing, deadening effect that repetition has on strong emotion. When first exposed to it, most people instinctively recoil from blood and gore and depictions of bloody scenes, as our mirror neurons automatically map the injuries of others onto our own bodies, prompting horror and vicarious pain. Other aspects of the stress response are activated, including rises in heart rate and blood pressure. We tend to find injury and physical signs of illness or deformity disgusting—this sense of disgust probably helped our ancestors survive because staying away from the infected or wounded reduced exposure to contagious disease and, in situations of violence, possibly injury. Incidentally, we can see another aspect of the power of empathy when we overcome these initial reactions and nonetheless help the sick or injured.

Repeated exposure to blood and gore, however, decreases these physiological and emotional responses. This is helpful for doctors and others whose jobs put them in frequent contact with traumatic injury or death: if you are overcome by your emotional reactions to the pain of others or by disgust, you aren't going to be much use in the emergency room or at a crime scene. But this is also why doctors, police, and rescue workers are notorious for insensitivity: our emotions rely on this physiological response. If you don't work frequently and deliberately to compensate for its absence, you can become inured to pain and suffering to which you should be emotionally responsive. Callous cops and cold doctors are stereotypes in detective stories and medical dramas for good reason.

And indeed, a study of medical students found that young doctors became increasingly less empathetic over the course of their training. Male doctors in specialties like pediatrics and psychiatry began with higher levels of empathy than members of the general public—but over three years of training, these declined until they were comparable to normal population levels. For female doctors it was worse. They began with normal levels of empathy, but wound up having even less after three years in med school.[11]

Although exposure to gory videos and gaming certainly isn't as numbing as having these kinds of jobs, it clearly can have an anesthetic effect. Studies have found that people who are less emotionally affected by seeing violence are more likely to be violent themselves.[12] And many contexts in which video violence is seen or enacted have no role for compassion. In fact, some specifically reward the opposite.

Indeed, the military now takes advantage of video games in training soldiers to kill in combat. Though obviously, people can and do frequently behave in brutal and inhumane ways without any prompting, most people, most of the time, find killing others deeply repellent. In fact, some research has suggested that in many wars in the past, a majority of soldiers in combat did not fire their weapons at the enemy.

Part of the explanation can be seen in an experience described by George Orwell. He found himself unable to kill an enemy soldier whom he was fighting against in the Spanish civil war, despite having a clear opportunity. He had encountered the soldier stumbling and unprepared, holding his pants up with one hand. "I did not shoot partly because of that detail about the trousers," he wrote. "I had come here to shoot at 'fascists' but a man who is holding up his trousers isn't a 'Fascist,' he is visibly a fellow-creature, similar to yourself and you don't feel like shooting at him."[13] Orwell had involuntarily imagined himself in the other guy's position, seeing him as a real person with particular quirks and traits. Once he'd empathized, he couldn't kill.

This is also why killing at a distance is much easier: it, too, reduces empathy. It's much less difficult to push a button to drop a bomb on people than it is to shoot someone you can recognize as a real human being. Similarly, it's a lot easier to shoot someone than it is to stab him. Using language or physical barriers like hoods to create distance is one aspect of this—literally making the enemy faceless again reduces the odds that empathy and compassion will intercede. In fact, late in World War I, commanders had to actively prevent peace from breaking out at some points because soldiers in the trenches got to know the guys on the other side and made deals to "fight" in ways that wouldn't actually hurt one another, such as deliberately timing attacks to avoid harm, and shooting over people's heads.[14]

Consequently, boot camp is deliberately designed to break recruits down, promote group loyalty, dehumanize the enemy, and eradicate this compassion before battle as much as possible. It aims to highlight us-versus-them thinking, rather than reduce it. This is another reason why teen boot camps are not a good approach to stopping juvenile violence, addiction, or crime. Their intent is to decrease empathy rather than create it, which is not a desirable effect when part of the problem in the first place may have involved a lack of concern for others. Because of their ability to desensitize people to gore and reinforce habits that are useful in firefights, first-person shooter video games can also easily be used as part of a training regimen to reduce this inconvenient empathy.

SO DOES THIS mean that a kind teen from a good family who plays several hours a week of *Grand Theft Auto* is at risk for becoming a school shooter? Of course not. As you likely already expect, he'll almost certainly grow up to be a wonderful human being, because parents, peers, and community are far more powerful influences in his life than video games. But there may be subtle effects. For example, one study compared college students who were playing popular violent and nonviolent video games. The researchers staged a fight in the hallway outside the game room in which one person was said to be injured. The violent gamers took almost a minute longer to help.[15] Is this likely to matter most of the time? Again, probably not.

The problem is that in our world now, there are more and more of these small influences against empathy. Although each one on its own isn't huge, they can start to add up. For example, part of what makes parents protective against the deadening effects of these games is the messages they send their children about them and the time away from the games doing other things that they typically insist on. But when parents have increasingly limited time to spend with their children, and electronic entertainment and structured activities and school cut into it more and more, the chances to reinforce these values are limited.

Nonetheless, it's important to note that when parents do take time to discuss—not lecture, but genuinely discuss—why their values make them concerned about violence, youths tend to get the message, even if they argue strenuously in favor of the games and their content. Research shows that these kinds of conversations help children avoid being influenced by violent content—but that lecturing—and, for teens, bans or extremely strict limits on gaming—can backfire. Restrictions can work for younger children. However, for teens, the lure of the forbidden means that being too tough can increase interest in this type of content that would otherwise tend to decline.[16] A better strategy is to provide or help them find attractive alternative activities.

Youths who are already aggressive, however, may find justification for their actions in games and other violent media. Such children are more likely to come from troubled families in the first place, because both genetics and a traumatic environment can contribute to increased risk for aggression. This is yet another instance of a vicious cycle: the most vulnerable children are most attracted to the kind of content that can make their behavior worse. They are also the least protected from it by their families and communities, and so, although researchers estimate that only about 4–9 percent of the variance in violence between individuals is attributable to video exposure, it can make an already bad situation worse.[17]

Of course, screen time isn't always bad—there are many ways, even while engaging with violent content, that children can also connect to people. A teenager who has no friends at school might find deep connections through online gaming. Social networking sites can link us to distant relatives and friends with whom we might otherwise lose touch. These contacts and the emotions they engage are real. And when online social networks or games add to face-to-face relationships—rather than substitute for them—they can improve our relatedness and compassion.

But we do think the issue is worth thinking more deeply about than the current debates over video violence and screen time have allowed. In his classic *Bowling Alone: The Collapse and Revival of American Community,* sociologist Robert Putnam explored how American social ties and civic engagement have unraveled dramatically over the last five to six decades. This includes stunning declines in virtually every form of social and civic participation from voting rates to churchgoing to time spent entertaining friends and the proportion of money given to charity. As we noted earlier, participation in local face-to-face groups has fallen by more than half since the 1960s, and the amount of time spent socializing with friends or having family dinners has similarly declined. On nearly all measures of social life—from the frequency of dinner parties or dining out with friends to the number of friends and confidants we have to our closeness with our families—Americans tend to have fewer and lower-quality interactions with one another than our parents and grandparents did.[18]

What could account for this dramatic tear in our social fabric? Putnam looked at many variables, including the movement of women into the workforce, the sharp rise in divorce since the 1970s, people moving from cities to suburbs, all of which have had a significant impact on social life. However, television viewing has had an outsized effect, with the introduction of TV disproportionately cutting into social and family time and depressing voter participation. Simply by keeping us away from one another, isolated in our homes, television has made us less trusting. And the content of many programs has made this worse. For example, people who watch more TV news tended to blame things like crime and unemployment on individual failures, rather than on social issues.[19] Frequent TV viewing is also linked with depression.[20]

Putnam's research cannot tell us much about the impact of the more participatory Internet yet—and there have been heartening recent increases in volunteering by youth. Though people have predicted dire consequences from our successively screen-engaged populace, so far, they have not materialized. The age of violent video games has been accompanied by a drop—not a rise—in violent crime, so their real-world impact is still unclear. Some argue that this means that video violence actually substitutes for real-world violence, allowing angry people to take out their rage harmlessly and acting as a positive force. This kind of substitution—replacing real enemies with virtual ones, rather than doing so with friends—would be wonderful.

Unfortunately, research finds that anger management therapies—like hitting pillows or other attempts at "catharsis" or "getting it all out"—increase rage, rather than decrease it.[21] There is no reason to believe that this is not true with video games, too; the mind is not a boiler that traps anger and will explode if it is not somehow expressed. Instead, anger is an emotion prompted by situations, some of which act to reduce it, others to enhance it. Let's say you have an abusive boss. While imagining him as your target in a video-game gun battle may be satisfying, it won't make the anger stop. Only either confronting the situation by talking to him, changing jobs, or reframing the experience as one that is not personal and needn't generate anger can do that. It is simply not the case that anger is stored in the brain and needs expression to avoid causing harm, and there's no evidence that it sticks around after you genuinely change the way you view whatever provoked it or remove yourself from the situation and let it go. Expression isn't always necessary, contrary to popular belief.

It comes down to this. Given what we know about video violence, there is strong reason to believe that it can be pernicious. Even with "positive" content, the youngest children simply can't learn what they need at their stage of development from screens alone—and some supposedly "educational" content has unintended consequences. Brandon's case is evidence of the poverty of expression and experience that massive overexposure to television can produce. Infants and toddlers need touch, warmth, presence, things even the most interactive computer cannot provide. Babies cannot develop the foundations of empathy without frequent, nurturing contact with a few consistent caregivers. Social development requires multiple repeated face-to-face interactions.

As we've said before, the brain becomes what it does. And the conditions of modern life conspire against allowing children time and space to repeatedly practice the social skills necessary for the true development of empathy. School focuses almost exclusively on cognitive development. In fact, some elementary schools have no recess whatsoever—according to play researcher David Elkind, forty thousand American schools have eliminated it entirely in recent years,[22] and other research shows that just 12 percent of states require elementary schools to provide any free time.[23] Structured activities now fill much of the after-school and summer time that was previously left for free play, and the realities of work and school life often eliminate the opportunity for family meals. Even kindergartners now face standardized testing. And then there are those constantly beckoning televisions, video-game consoles, mobile devices, and the Internet.

We're not saying that parents should never park their children in front of a screen or allow them to zone out from time to time. Parents' sanity and ability to be present also relies on respite and their own need for occasional solitude and escape! We're also not suggesting we go back to the caves—in fact, as we'll see, there's some intriguing evidence, if you take a very long view of history, that empathy may actually have increased over the course of centuries. But we are arguing that we need to be much more conscious of the intrusions of various media into our lives and the drip-drip-drip of various trends that together can spell trouble for our ability to trust and connect.

eleven | ON BABOONS, BRITISH CIVIL
SERVANTS, AND THE OSCARS

W E'VE ALL HEARD THE ADVICE FOR living longer and staying healthy: don't smoke, eat your fruits and veggies, exercise. Here's another tip, though not many of us can manage it. Win an Oscar. A recent study found that Academy Award winners lived, on average, four years longer than those who were happy just to have been nominated. Unfortunately, this life-extension opportunity applies only to actors: screenwriters apparently do not gain longevity if they win the otherwise-coveted golden statuette. For actors, however, the impact of winning an Oscar is similar in terms of life-extending power to having their risk of dying from the number one cause of death, heart disease, reduced to zero.[1]

And herein lies a tale, one that connects what we've seen so far about how relationships are critical to health and how social influences on our stress response system can be either panaceas or poisons, depending on their direction. As we've seen, we are all much more dependent on one another for physical and mental health than researchers ever suspected—and one way this works is through our experience of social status. To understand how empathy threads itself throughout our relational interactions and profoundly influences health, we need to explore the roots and results of leadership and social hierarchy. This will take us on a wide-ranging journey: from Bruce's work with cults like the Branch Davidians and the Fundamentalist Church of Jesus Christ of Latter Day Saints (FLDS) to the halls of the British Civil Service and a visit to a unique troop of mellow baboons.

"IT WAS MY blessing. I was very happy." The young mother, Brenda,* described her wedding day. She smiled and touched her hair self-consciously. It was blond, elevated into a "wave" in the front and elaborately French-braided in the back.

"My father asked me to come with him," she continued.

"And when you left home that day did you know that you were going to get married?" Bruce asked.

"No. When we got there, he told me the Prophet had decided that it was time for me to marry and that William* would be my husband."

"Did you know him before that day?"

"No."

"How did you know you would like him?"

"I trusted my father and the Prophet. I was very excited."

Brenda was sincere. Bruce strongly felt it. Though he knew she wouldn't admit it, she had likely been thirteen on her wedding day. Her husband was in his forties. And she was far from his only wife.

"And you got married that day?"

"Yes."

"Was there a ceremony?"

"Yes, we were united for time and eternity."

"Did the rest of your family from Utah come?"

"Well, no. They didn't," she admitted, and Bruce could sense her shutting down. "I don't want to talk about sacred things," she said.

Bruce was sitting in a small cinder-block office in San Angelo, Texas. He was talking with one of the young mothers from the Yearning for Zion Ranch in Eldorado. That eerie compound—with its imposingly blank all-white Temple and similarly spare living quarters—was then inhabited by about seven to eight hundred members of the FLDS. The FLDS is a splinter group of Mormon polygamists who believe that the main church was wrong when it abandoned polygamy. The members who had moved here from the group's home base on the Utah/Arizona border were the most obedient and faithful, as determined by their "prophet."

The conversation took place only days after a high-profile raid by Texas Child Protective Services (CPS) in April 2008. The agency had removed about 430 children from the ranch—in what is believed to be the largest child custody action in history. A battered women's shelter had received urgent calls for help (now widely seen as a hoax orchestrated by a former member). The caller claimed to be a teenage girl who was being sexually abused there. Viewing the whole compound as a single household, state authorities decided that all of the children were at risk and removed them.

Bruce was serving as a consultant to help the state determine what was best for the children. What he has learned from studying the FLDS and other cultlike groups can help us understand some of the most important and frequently overlooked influences on empathy and human relationships: power and leadership. By looking at these extreme situations, we can also start to understand how these hierarchies affect our health and the health of whole societies.

Over the years Bruce had developed some expertise working with members of cults and cultlike organizations. It started in 1993 in Waco, Texas. Another isolated religious sect, the Branch Davidians, had stockpiled weapons and was also believed to be abusing children in its fortresslike compound. Seeking to arrest leader David Koresh, the federal bureau of Alcohol, Tobacco and Firearms (ATF) conducted a disastrous raid, in which four agents and six sect members were killed. A six-week standoff ensued, culminating in the tragic fire that killed eighty followers of Koresh, including twenty-three children.

Bruce led the multi-organizational clinical team that worked with twenty-one children released early in the siege through negotiations with the FBI. From that time forward, Bruce has met and worked with hundreds of other children and adults who grew up in these kinds of groups, ranging from isolated, patriarchal family groups with multi-generational histories of abuse to well-organized international cults such as the Children of God.

These organizations are overwhelmingly authoritarian. The leader defines the values, behaviors, and emotional climate. Their hierarchical structure makes him—and it is almost always a man—a major determinant of the emotional and even physical health of the group. If the leader is centered, compassionate, dedicated, and generally healthy, the group will reflect this. If the individual who takes this power is sociopathic, sadistic, violent, racist, or filled with other hateful beliefs, the group can easily be bent to his will.

Bruce had watched Brenda come into the room. He rose and extended his hand. She was small, young, and girlish looking. She claimed she was eighteen. She was the mother of a toddler. She passively held her arm out, with a limp hand, and turned her face toward Bruce. She did not make eye contact. He thanked her for agreeing to speak to him and explained his role, emphasizing that she didn't have to answer any questions she didn't want to and could leave at any time. She nodded at all of this in polite compliance.

The interview continued with her partial responses spoken in a quiet, childlike voice. She seemed oddly robotic; compliant and cheerful, yet withholding. It reminded Bruce of a similar interview he'd conducted with a Branch Davidian mother fifteen years earlier.

Paula* was older, but she shared the timid, congenial, and deferential demeanor of this much younger FLDS mother. She lived at Ranch

Apocalypse with her husband and five children. At the time of the ATF raid, her whole family was in the compound. In the first two days, her three youngest children were released to the FBI and put in the care of Bruce's clinical team.

As the children were shuffled into an armored car in the middle of the night, one of the Davidian mothers kissed her child and said, "Here are the people who will kill us. I will see you in heaven." Unfortunately, she was, in some sense, right. Halfway through the siege, several mothers took advantage of an opportunity to leave the compound. Paula was one of these few survivors. But her husband and older children stayed behind—and died in the fire several weeks later. The day after the inferno, Bruce met with Paula to talk about how she was addressing her own grief and loss—and her children's. The interview was stunning.

Paula was smiling and calm. She was open about her continuing belief in David Koresh as the messiah. "David said this would happen. It is the fulfillment of his prophecy," she told Bruce. She was beaming, seemingly happy. The only sorrow she expressed was that she hadn't been among those who were already in heaven—but thankfully, Bruce didn't sense that she would take any suicidal action to "reunite" her family. She was too passive; she wouldn't make that kind of a decision without a leader's direct command.

Bruce learned from Paula, other adult members, ex-members, and the children he worked with about the history of the Branch Davidians. Koresh had gained control of a strictly hierarchical sect that had a long history of relatively ordinary fundamentalist Christian beliefs. He was a charismatic, articulate, bright but very troubled man. His mixture of passage-selective, self-serving apocalyptic theology fused with the existing group's vulnerability to authoritarian control. Slowly he started to change the group's core beliefs. In each case, when Koresh presented new rules, he said they were revelations from

God. He used several well-known tactics to convince his followers of their truth. The first was repetition—he would repeat his points over and over and over, until followers were so dulled that they could no longer distinguish between his thoughts on the subject and their own. The second was rhythm. His preaching was often conducted in soothing cadences, well-timed sermons that preyed upon those associations people have between certain rhythms and the safety of the womb. Both of these tactics also relied on individuals' natural mirroring responses to connect and unite the group under his leadership.

And he used fear and discomfort to create vulnerability. Although he sometimes preached for hours, followers were not allowed to leave for food, bathroom breaks, or sleep. Even these small discomforts move people higher into the arousal continuum, and this, as we've seen, makes them more vulnerable to persuasion by reducing and sometimes even shutting down rational thought in the cortex. With fiery, apocalyptic preaching—and attendant shouting and expressions of rage—he added to these small tastes of anxiety real, gut-wrenching fear as well. Neither adults nor children wanted to face his wrath— which could include beatings, food deprivation, or shunning and isolation within the group.

However, by being the person who could relieve their discomfort, Koresh became a source of reward, too. Cult leaders manipulate the stress response system in a pathological replay of a mother's modulation of her child's stress. The trick here is that they deliberately induce distress—so that when they relieve it, they will also be the source of your pleasure. This leads to a powerful and, to outside observers, puzzling connection between cult leader and cult member. The same thing can be seen in abusive relationships and in "Stockholm syndrome," where crime victims fall in love with or become supportive of their captors.

Over time, Koresh's intense style led all but the most committed members to leave. By the time the group built and occupied what he called "Ranch Apocalypse," the remaining Davidians were all vulnerable to his persuasive demagoguery. Now it took little for them to accept his further "revelations." First, he segregated men from women, husband from wife. Earlier, he merely controlled couples' sexual relationships—determining the frequency and timing. Then, he banned sex entirely. His next revelation was that he should be the father of all the children—both existing and in the future. He began having sex with all of the women, including very young girls. He even asked the parents to present their daughters to him at puberty. This was seen as a mark of high status, an honor. Young girls eagerly awaited the chance to become a "bride of David"—they saw this as literally being married to God.

Similar echoes of their leader's behavior and similar controlling tactics could be seen in the FLDS. Marriage to older, high-status men—especially the prophet himself—was an honor for young girls. The state of Texas wasn't concerned about polygamy between consenting adults—it was polygamous marriages of underage girls who had been raised to believe that they had no other choice that were being targeted. Brent Jeffs, a nephew of the group's incarcerated prophet Warren Jeffs and grandson of the previous prophet Rulon Jeffs, met with Bruce in San Angelo. Brent had been instrumental in the downfall of his uncle, who had raped him and several of his brothers when they were young children. It was Brent's civil suit related to that abuse that had sent Warren underground, where he became one of the FBI's "most wanted."

Brent and his family—his father, mother, her two "sister-wives" (one of whom was his mother's biological sister), and his nineteen brothers and sisters—knew all too well how the health of the FLDS reflected the health of its leaders. When Brent's parents were first married, their prophet had been "Uncle Roy," a genial man who felt that the group should be relatively open. He seemed to take women's preferences into account when arranging unions. When they felt ready to get married, women would "present themselves" to him and he would ask if they'd had any "revelations" about who their husband should be. Young girls weren't simply assigned at the prophet's whim. The group was strange, but nowhere near as unhealthy as it later became.

When Brent's grandfather Rulon took over after Roy's death, the FLDS became increasingly isolated and fear-driven, like the Davidians. As Rulon became increasingly incapacitated by strokes, his son Warren took over. Both preached about the impending apocalypse—and how the people needed to become "pure" to prepare to be lifted up to heaven. They used the same kind of repetitive, lulling preaching alternating with the discomfort and fear that Koresh had used to instill obedience. Nonreligious books, movies, and all TV except for news were banned. Schoolchildren were taught survival skills—like slaughtering animals and eating insects, complete with bloody demonstrations.

When Rulon died in 2002 and Warren took over, the community became even more closed. Now pretty much the only "entertainment" people were allowed was listening to tapes of Warren's preaching. He arranged all of the marriages himself—and began assigning girls to men at younger ages. If a man displeased him, he frequently "reassigned" his wives and children to someone else. Afterward, the father would have no contact with his family; often, he wouldn't even be told where they lived. Fear of this prospect produced more obedience.

One of Warren's first moves upon becoming prophet was to take for himself his father's youngest and prettiest wives. When he was convicted in 2007 as an accomplice to rape for forcing fourteen-year-old Elissa Wall to marry her nineteen-year-old cousin, he was believed to have at least forty wives. The leadership vacuum caused by Warren

Jeffs's five-years-to-life sentence leaves room for the group to change once again: in whatever direction the new leadership takes it.

SO WHY DO so few members leave these groups voluntarily—and why do they accept such harsh conditions? Both the Davidian Paula and the FLDS member Brenda were sincere, decent people. They both worked hard. They were both loving mothers: indeed, the FLDS community in particular included many healthy elements for children, like rich opportunities for play, interaction with multiple generations, and a way of life that prioritized family. Both women had strong beliefs in their values. Ultimately, however, both had been exploited by their "leaders." Why do we follow the leader?

Of course, we already know how important families and friends are to human health—and that this importance is rooted in the way social connection modulates the brain's stress response system. We know, too, that the brain works this way because humans evolved in small groups. But because humans are our own best friends and worst enemies, this sad double-edged human nature creates a need for our brain to continually appraise and categorize our relational milieu. Are we in a safe, unknown, or dangerous situation, human-relations-wise? Our brains are always measuring the tone of our emotional environment.

Two key factors in this appraisal are our knowledge of our place in the hierarchy and our sense that people are "safe and familiar." When we know everyone and share pretty much the same status, this doesn't cause too much stress. But as humans moved from living in settings where status differences were minimal to more complex, hierarchical societies with more strangers, this task became more demanding. Still, we need to know where we stand. Figuring out who has power and who doesn't is a critical part of staying safe. Our stress response and relational systems are cued to these power dynamics—consequently, our social status in hierarchies plays a big role in our health. Leaders like Warren Jeffs and David Koresh exploit features of the stress response to produce compliance and obedience. One of the best ways to do this is to create—as both of them did—a climate of unpredictability and pervasive threat.

The higher level of arousal created by threat pushes the brain to rely on lower, less thoughtful regions. Thus, being in a higher alarm state makes people more vulnerable to manipulation. It also makes them more likely to be compliant. As with so many other characteristics, this trait was probably adaptive in early human life but is less so now. During an emergency, it's not useful for people to fight over who's in charge. This wastes essential seconds that could be used to coordinate an effective response. Consequently, all but the most dominant people tend to become suggestible and passive under threat. This tendency is useful during true catastrophes: those who have survived plane crashes or natural disasters often find themselves surprised by how cooperative and almost eerily calm the people around them seem. This is a result of this response to severe stress.

Though we tend to see groups as vulnerable to panic in such situations, crowd panics are the exception, not the rule.[2] Rather than trampling one another on the way to the exits, people instead tend to do as they are told by those who take charge. If no one leads, in fact, people can lose precious time needed to get to safety by waiting in this passive, bewildered state. Here is another case where it is clear that nurture and altruism are in our nature: under extreme threat, cooperation is actually the default response for most of us.

For people like Brenda and Paula, however—under the sway of ill-intentioned leaders—this aspect of the stress response makes them vulnerable. The compliance reaction doesn't just suddenly take hold under extreme threat. It applies to a greater or lesser extent during sustained low- to moderate-level threat as well. The lower people are on a dominance hierarchy, the more susceptible they are to becoming compliant when stressed. It is easier to intimidate, threaten, and essentially immobilize such people. And the greater the threat, the more mild, meek, and easy to lead people become because they are also less capable in this brain state of rational or abstract cognition.

These kinds of responses account for a lot of the otherwise inexplicable behavior in which some cult members engage. Of course, this doesn't mean that cult members have no responsibility for their actions. It's just another example of the effect that our internal response to stress has on our behavior. It's kind of like what happens to people high on alcohol or other drugs. When smoking crack, someone who is already inclined toward aggression may become more violent—but someone who is typically timid will tend to become even more cautious and withdrawn. Usual tendencies are exaggerated by situations, rarely created from whole cloth. Over time, however, cults—and addictions, for that matter—can gradually shift people's values in directions that seem unrecognizable compared to their initial starting point.

Even in the absence of an obvious threat to personal safety, however, our brains are always looking for information about the human social economy. These are related to timeless issues in our relationships: friend or foe, mate or no, ally or rival? The parts of the brain processing these ongoing issues are focused on the present moment. Some element of planning can be involved, but the capacity to read nonverbal cues is mediated by a primitive portion of the brain—a precortical region that has little sense of the future. This means that responses that result from this processing—in the absence of input from higher brain regions—will be focused on short-term outcomes. The signals, cues, and patterns of the moment are being matched against stored memory templates of similar experience in order to answer the core question of all interaction: Right now, should I cooperate or compete?

During any interaction, the people involved are trying to read these cues. This continuous monitoring of power, threat, and status creates perceived differences in power. Someone will feel more dominant, someone else, more vulnerable. The difference between the position of power and control when any two people interact can be called the "power differential." This differential can wax and wane depending upon the situation. You can feel dominant at home with your children but vulnerable when having an annual review from a supervisor at work. The power differential can shift dramatically between two people in different contexts. When the boss in his three-piece suit dresses down a junior associate in a meeting, there's a very different power differential than when that same associate strikes the boss out at a company softball game, the boss wearing his plaid shorts and dorky calf-high socks.

A number of universal factors can also trigger power differentials. One of the most basic is size. When the brain has to establish upward gaze to enable eye contact, this triggers an automatic perception of mild vulnerability. This is an obvious issue when a six-foot-tall adult tries to talk to a tiny five-year-old. And it may be part of why the tallest presidential candidate usually wins. Attempts to enhance or minimize status differences by using this perceptual bias are pervasive. For example, in the courtroom, the judge is seated higher than everyone else, and the jury is elevated above the spectators, attorneys, and defendant. In the office, bosses often make sure that they have the most elevated position; visitors sit across from them in smaller, lower chairs. To do therapy with a child and create a sense of safety for the child, a good clinician will get down on the floor—at eye level with the child. Another obvious trigger for a power differential is having a weapon. The guy wielding the gun—no matter how small he is and how puny his muscles may be—is at the top of the power differential. The universal sign of nonthreatening greeting involves demonstrating that you bear no weapon by raising one or both empty hands.

But some of the most important determinants of these power differentials are culture specific. For example, British people can instantly distinguish between prince and pauper by hearing a countryman's accent, whereas this information would bypass most American listeners. Clothing and dress cues are similar, particularly uniforms that signify authority. For most people, talking to a uniformed police officer—even when we have done nothing wrong—can make us feel vulnerable. The trappings of authority will move us higher on the arousal continuum, making us more obedient and easier to control. The number of rapes and abductions that have been facilitated by using uniforms to gain initial trust or entry is testament to this.

The key to understanding the power of leadership—the answer as to why we follow the leader—lies in this power differential created between leaders and their followers. If leaders are wise and benevolent and have true competence in the area in which they are leading, they will not exploit this differential. If they are pathological, corrupt, or incompetent, the group and its members may be in big trouble. Cult leaders—and sociopaths—are masters at using the power differential to further their own goals and needs.

The power differential also has a large impact on the health of whole countries and populations—and this is where we get back to the Oscar connection. The more highly stratified a society is, the worse the health of that society is overall—especially for those on the bottom of the totem pole, but surprisingly, for those at the top as well. The lower you rank in such a society—the more often you are on the bottom of the power differential and the more unpredictable that is—the more likely you are to have health problems. The higher your social status—even in Hollywood—the healthier you will tend to be. But in cultures or social systems that are more equitable, everyone is better off.

So how could health be connected to these facts about leadership and why would empathy matter here? Once again, it's related to that social modulation of the stress response system. If you wind up on the bottom during most of your daily power interactions, you are, by definition, under chronic stress. If each encounter with other people ends with you as the one in an "alarm" state, you will literally wear out your body.

The pioneering research on social status and health was done in the United Kingdom, examining civil servants whose rank in the government bureaucracy is easy to track. Because Great Britain has national health insurance, the connection between social status and health there cannot be due primarily to lack of access to health care among the lower-ranking bureaucrats—all are covered by the same system, though the wealthy can add private insurance if they so desire. In this study, those on the bottom are not truly poor, simply working or lower middle class.

The "Whitehall Study"—and a second study that included female civil servants as well as males called "Whitehall II"—found a massive "social gradient" in health. This wasn't just one correlation: it was a clear slope downward stepwise to worse health from top to bottom. The first Whitehall study—so called because this is the name of the headquarters of the British Civil Service—included eighteen thousand men. It found that at ages forty to sixty-four, the lowest-ranking men were *four times more likely to die* from any cause than those who'd reached the top of the greasy pole. The gradient directly tracked a man's rank in the bureaucracy, so those of the second rank did better than those of the third, and the top rank was best of all.[3]

Other studies—for example, those looking at a country's population as a whole—have typically found a gradient where those at the bottom have twice to three times the mortality risk of those at the top.[4] Studies that include women (like Whitehall II) tend to find interesting

differences in status-related risk, where a woman's health risks are
more correlated with the overall income level in her household than
with her own job status.[5]

What's going on here? Isn't being the boss stressful, too—and
don't managers have more heart attacks than sales assistants? It turns
out that the impact of stress on health—and the way it is affected by
hierarchies—is most related to the amount of control you have over
your circumstances and work. What's most stressful is not being in
charge and taking responsibility for big decisions—but instead, being
held accountable for outcomes over which you have little or no control.

Although some high-level jobs are characterized by high accountability and low control—president of the United States comes to
mind—for the most part, such a squeeze tends to occur lower down in
the ranks. Think billions in bonuses for Wall Street executives who
ruined the economy—and layoffs for workers whose actions had nothing to do with what went wrong. Executives have great control and,
often, reduced accountability for failure; their employees have much
less control and are subject to much more severe consequences. Further, the less status you have, the less likely you are to have choices
about the type of work that you do or the hours when you do it. You
have less access to work that would be particularly meaningful to you,
and there is reduced choice and control over virtually every aspect of
your life. Meaning, choice, and at least some degree of control greatly
mitigate the stress of top positions like president.

Typically, then, being on top is a lot less stressful than being in the
middle—but even being in the middle is better than being on the bottom. The idea that executives are at greatest risk for heart disease
turns out to be false. However, there are situations where the stress of
seeking higher status does have negative health effects, which we will
explore later.

For now, we need to go back to the brain, to our old friend the stress response system, to really understand how chronic, uncontrollable stress could cause these negative effects on health. Consider your childhood—if Dad or Mom came home in a bad mood, how did you respond? How did you feel? Or think about work: most of us have seen a bad mood flow downward from the boss to everyone else in the office or shop. This flow of mood contagion from the top to the bottom takes place even among baboons, though they tend to play it out more physically than humans do. And we can start to understand how and why chronic, uncontrollable stress is so bad for our health by looking at some research on these animals.

Robert Sapolsky, who has studied stress among baboons on the African savannah, described one relevant status interaction this way: "A middle ranking male gets trounced in a fight, turns and chases a sub-adult male who lunges at an adult female, who bites a juvenile, who slaps an infant."[6] If baboons had pets, the dog would get the next kick. Sapolsky studied the effects of rank on stress hormones among these animals both living wild in the Serengeti and in controlled experiments. The troops of animals he studied in Kenya lived in a virtual baboon paradise: there were few predators and plentiful food, so the primates only had to spend about four hours a day "working" to stay well fed. Consequently, most of the stress they experienced was related to social issues.

To study their stress hormones, Sapolsky had to develop a technique in which he briefly anesthetized each animal at the same time each day by shooting him with a dart from a blowgun. Then he'd sneak over and take a quick blood sample. He had to keep his intervention brief and as minimal as possible. Otherwise, all that he would discover was that suddenly being shot out of nowhere by an anesthetic dart from a blowgun is stressful or that being attacked by a baboon causes great stress in humans! Because the females were often pregnant or nursing, and this technique could potentially harm them, he looked primarily at males. He took samples only on days when they had not engaged in fights or other activities that would be an obvious explanation for elevated stress hormones.

His findings paralleled the Whitehall study in many ways. Like lower-ranking humans, low-level baboons had elevated levels of stress hormones. "Good" HDL cholesterol was lowered in both bottom bureaucrat and baboon, which is not healthy. Among the low, blood pressure was high. And their immune systems were sluggish, as was their response to further stress. In other words, the baboons had all the precursors to disorders like heart attack, stroke, infectious disease, and other conditions that can explain why high-ranking humans are healthier and longer-lived than those with less status.

Since baboons don't smoke or drink, the social gradient here couldn't even partially be explained by bad behavior. (In the Whitehall study, only about one-third of the social gradient in health could be explained by higher rates of smoking among the lower ranked—however, even the smokers showed a gradient favoring the high status.)[7] But could these signs of ill-health be what caused these animals—or the humans, for that matter—to have lower rank? After all, sickly animals aren't likely to dominate. Lab studies can answer this question in animals by creating groups artificially and seeing who comes out on top. For baboons, it turns out that being lower ranked produces the stress-related decline in health, rather than the other way around; and research suggests that this is also true for humans.

Interestingly, however, not all primates have status-linked stress issues. Low-ranking marmosets—adorable, six- to seven-inch-tall primates found in South America—do not seem to suffer from worse health than dominants. Being a low-ranking marmoset is less stressful than being a low-ranking baboon. Low-ranking marmosets don't get harassed or hit out of nowhere by their superiors. They don't have to deal with this type of bullying or threat from above. Their role is to babysit for the offspring of their older siblings and cousins; in fact, the dominant female can literally turn off the reproductive hormones of the lower-ranked animals, rendering them infertile. Since they share genes with these offspring and can later become dominant, this isn't as bad a deal as it would first appear to be. Interestingly, marmosets are cooperative breeders, where the parents aren't the only ones who are important to children's survival. As noted earlier, anthropologist Sarah Hrdy has argued that this is how humans should probably be classified as well—and this offers another hint as to some probable characteristics of a healthy society, as we'll see in the next chapter.

In contrast, animals that show a large stress response to hierarchy are typically species in which dominants harass and intimidate those in the lower ranks, by doing things like taking the best food from them and even attacking them physically to demonstrate superiority. Such species also tend to have harsh penalties for lower-ranked animals that get out of line. For example, if a subordinate baboon gets caught trying to mate with a female above his station or refuses to give up food demanded by his superior, the punishment can be a vicious slash from the dominant's sharp canine teeth.

In some animals, however, it is indeed the higher-ranking individuals that show the elevated stress hormone profile. If you think about the importance of control over stress as a variable, you can probably guess which situations produce high stress at the top. Indeed, this is seen in species in which rank is highly unstable and the betas are always plotting coups against the alpha, for example wild dogs and dwarf mongooses.[8] Instability of rank can be as great a source of stress as low rank itself. It's terrifying not knowing where you stand. Or, as the great family therapist Virginia Satir put it regarding human life, "Most people prefer the certainty of misery to the misery of uncertainty."

So our stress systems are mediated by those around us, and they

are sensitive to rank and status. There are obviously individual tem-
peramental factors that play a role as well. These can be seen even
among Sapolsky's baboons. Take "Kenneth," a high-ranking male,
whom the researcher observed achieving that status by "playing nice":
grooming friends, making allies, and defusing conflicts. Kenneth
once got a chance to be the alpha, but he decided not to fight it out
with the big guy after he'd risen to become number two. He was happy
where he was. As many an insincere executive resignation letter
reads, in this case, he actually did decide to spend more time with his
family rather than be CEO. When Sapolsky darted and studied him,
he found that Kenneth's stress response wasn't much affected by status fights—he got by with help from his friends. Even in a status-
obsessed world, Kenneth had a haven in others. "Gary," on the other
hand, fought his way into the number two spot in the troop at one
point. Fighting and power plays—not cooperation—were his specialty.
He was always on guard—and didn't back down when threatened.
Consequently, Gary was always on edge, even when high in rank. We
can all think of human examples of both strategies.

Does this mean that humans as a species are doomed to status-
linked disease? We certainly do have hierarchies, status obsession,
and deprivation and harsh punishment for those on the bottom. Even
the highly egalitarian Scandinavian countries haven't eliminated the
social gradient in health—but notably, they have reduced it significantly. And, in fact, even Sapolsky's baboons aren't genetically stuck
on status-linked severe stress, which means that highly flexible humans are not likely to be so, either. An accidental or, as scientists call
it, "natural" experiment found a sharp change in the stress profile of
one troop Sapolsky studied following an outbreak of tuberculosis.

In the early 1980s, the "Forest Troop" baboons slept in trees within
walking distance of a resort—one that wasn't the "eco-friendly" type

seen in many nature preserves today. Out back, the tourist lodge had an open garbage pit. Around this time, the pit greatly expanded, soon becoming the major source of food for another troop whose home base was located much closer to the hotel. The "Garbage Dump" troop was, not surprisingly, territorial about access to this rich food resource. Consequently, only the most aggressive males from the Forest Troop were willing or able to feast at the dump patrolled by the other troop.

Unfortunately, in 1983, infected beef left at the dump caused an outbreak of bovine tuberculosis, which can kill baboons as well as cows. By 1986, the Garbage Dump group was wiped out—and the aggressive males from the Forest Troop who had eaten garbage were dying off rapidly as well. About half the males in the Forest Troop had been aggressive enough to eat at the dump. Their deaths ultimately left the group with twice as many females as males, and by 1993, none of the males who had lived through the epidemic was alive.

With so few males and so many females, competition for mates among males was dramatically decreased—and you would expect that that alone would reduce aggression. But that wasn't the change that Sapolsky found. The number of dominance interactions among males remained the same—and the reigning alpha kept his hold on power for the same amount of time, before and after. Status fights still existed, power plays still occurred. However, aggression by high-level males against those more than two ranks lower—the kick-the-dog kind of aggression that isn't a real contest over status, but simply bullying by a guy who knows he's going to win—dropped dramatically. There was also more grooming of males by females and vice versa—and males spent more time around females and juveniles. Basically, the Forest Troop became a place where, even for the lowest-ranking males, life was pretty good.

And so, bullying decreased—and affection rose. And this more peaceful and egalitarian lifestyle was visible in the baboons' stress hormones and accompanying biochemistry and in their reaction to an anxiety-provoking drug.[9] Lower-ranking males in the Forest Troop after the epidemic did not have the stress-related pathology of their fathers. This more easygoing way of life seems to have been maintained in this group, even as other nearby troops retained a less equitable "culture."

Of course, human males might get nicer, too, if there were two women for every man! Women do tend to be less physically aggressive than men and a female-dominated world might well include less violence—though one can imagine women becoming more competitive with such a skewed sex ratio. However, Sapolsky compared the Forest Troop to other troops with unbalanced gender ratios favoring females and they did not show these behavior patterns. Some kind of genuine cultural change seems to have taken place. It would be fascinating to see how females nurture their offspring in this troop and if differences there are linked with this kinder, gentler society. We would bet that they are.

From other research such as the ACE studies, we now also know that changes in the stress response systems originating in childhood can play a role in these status-related health problems in humans. The Whitehall studies found that people from poor or working-class backgrounds who rose in status as adults were more healthy than those who stayed in those classes—but less healthy than those who had been advantaged for a lifetime. The effects of stress make their mark early, because stress modulation is learned in the cradle and even in the womb, as maternal stress hormones affect the development of infant stress systems. As we saw in Trinity's case, her weight and related health problems are rooted in her stressful childhood.

BUT HOW DOES this happen? Why is uncontrollable chronic stress so bad for the brain and body in the long run? Understanding this makes it even clearer why empathy and relationships are critical to human survival. We've seen how the stress response system is activated to power fight or flight, sending stress hormones everywhere and raising heart rate and blood pressure; we know how affection can reverse these effects. In modern life, however, situations in which fighting or fleeing is appropriate are few and far between—and, simultaneously, we have fewer opportunities for calming human contact. Nonetheless, activation of this stress response is not rare. Loud noises, strange faces, looming deadlines, bills, arguments—things that produce some degree of tension are everywhere. The bottom line is that our stress response system evolved to handle a very different set of stressors than the ones we face now.

And while activating this system in the short bursts for which it was designed can save your life, if it stays active for too long, it is harmful. Take blood pressure. High blood pressure is useful for fight or flight: this gets blood to the muscles and other areas where it is needed to maximize speed and strength. However, long-term high blood pressure is dangerous, straining blood vessels throughout the body.

This is made even worse by the fact that the chemistry of chronic, elevated stress hardens and clogs arteries—meaning that you are essentially putting more pressure on stiffer, smaller pipes. When blood vessels clog or spring leaks in the brain, this is called either "ischemic" (blockage) or "hemorrhagic" (bleeding) stroke. Neither is good. And the plaques or clots that fur the arteries themselves can damage the heart, producing heart attack or other potentially deadly cardiac problems. Alternatively, these clots can travel to other areas like the lungs and kill you by something known as pulmonary embolism.

Add in smoking and overeating—which make atherosclerosis and high blood pressure even worse—and you can easily see why stress can, over time, seriously affect risk for the number one cause of death in America: cardiovascular disease. Since people lower in social status experience more stress as a result of their position in the world, one obvious consequence is an increased risk of heart attack and stroke—before they even think about drinking, smoking, or taking drugs to try to cope, in fact, before they even leave the womb. The mind/body connection and physical effects of chronic, uncontrollable stress on health could not be more clear.

Raising blood pressure and gumming up arteries are not the only ways elevated stress hormones can damage physical health, however. Note that our low-status baboons and Brits also suffered impaired immune function. This is because, once again, when you are fleeing a bear, it's not useful to devote precious energy to keeping the immune system at top form. If you've ever had an allergy or a condition like asthma or psoriasis, you might have been given prednisone—or another kind of drug in a class of substances called "glucocorticoids." These steroid hormones are naturally produced during the stress response, and they lower immunity. In medicine, this property is useful to tune down the immune system when it overreacts. In the long term, however, glucocorticoids can have serious side effects because suppressing the immune system quite obviously produces problems of its own.

As a result, chronic stress can reduce the ability of your immune system to mount a response to everything from infectious diseases like tuberculosis and flu to cancer. Like heart disease, these illnesses help create the social gradient in health—though there are certain cancers that show an opposite gradient, affecting mainly more affluent people. Why this is the case is not known—but, as a whole, they do not have enough of an impact to make it more healthy to be poor than rich.

To top it all off, high chronic stress that starts in early life doesn't just cause physical illness. It increases the risk for virtually every mental illness known—including bipolar disorder, schizophrenia, addiction, and obviously, posttraumatic stress disorder.[10] This makes sense when you consider how widespread the influence of the stress response system is on the brain and the connections between its key neural networks and every single brain region. When it is abnormally and persistently activated by stress, its impact will be pervasive. Persistent stress will alter the biology of the brain.

One of the best-characterized probable mechanisms involves some forms of depression. Here's what likely happens. Chronically elevated glucocorticoids (a result of stress) are known to be dangerous to certain regions of the brain. By causing neurons to release too much of a neurotransmitter called "glutamate," these substances literally kill brain cells in the hippocampus, a region important in multiple functions including memory. This damage could be one reason depressed people can have memory and concentration problems. Many other systems are involved as well, and this is only one possible cause, of course.

But studies show that some people with depression have measurable shrinkage of the hippocampus. Further, almost everything that successfully lifts depression seems to ultimately produce regrowth of cells in that region. Whether it's talk therapy, electroconvulsive therapy, or medication—recovery often involves repairing the hippocampus. So, by damaging this region, high stress can exacerbate or cause depression—and by repairing it, these treatments can help.[11]

Given the ubiquitous reach of the neural networks involved in the stress response, it is also not surprising that there is a strong correlation between depression and heart disease: the same stress increases risk for both. Through a number of potential mechanisms, prolonged stress can affect the physiology of the heart. The head and the heart are not far apart at all—and part of what heals the heart both metaphorically and literally is affection. Heartbreak and loneliness are physically nearly inseparable. "Broken heart syndrome"—in which spouses die in quick succession—is the most poignant example. Loss of close relationships is not just metaphorically heartbreaking.

There's another way that status stress may be linked with depression. Some have argued that depression is, in essence, the chemistry of being subordinate, that it is often triggered by declines in status or loss of social support and that its postures and sense of resignation in humans are what being a bottom baboon "looks like."[12] This clearly cannot account for the entire range and complexity of human depressive symptoms—but the endocrine profile of some depression is similar to the endocrine profile of low-ranking social mammals. It involves chronically elevated stress hormones. Further, if you boost levels of the neurotransmitter serotonin—higher levels of which are linked to relief of depression—in vervet monkeys by giving Prozac or similar drugs, you can raise the animal's social standing. If you deplete serotonin, rank falls.[13]

From this perspective, some traits that may predispose to depression can be seen as an evolutionary adaptation for social living. Since not everyone can be the alpha or beta, some mechanism is needed to prevent subordinate animals from constantly fighting to try to raise their status. Seen in this light, subordination is a way to keep us from fighting unwinnable battles, to make us withdraw and behave in a way that signals acceptance of defeat to the winners. In this way, it minimizes injury. Whether this line of research has some clues to the mechanisms underlying depression remains to be seen, but it provides a plausible and promising direction for exploring depression-related traits.

This role of depression as a defeat signal may also explain part of why having control over a situation tends to protect against depression and other harmful effects of ongoing stress—and why lack of control has such deleterious physiological consequences. This is true even in animals: a sense of control can make the difference between stress that damages health and stress that promotes it.

These connections—and their relationship to the role of stress in depression—were discovered almost by accident, as is the way in a lot of science. University of Pennsylvania psychologist Martin Seligman and his colleagues were studying learning in dogs, trying to teach them to associate a sound with a shock. Some of the dogs could escape the shock by pressing a lever; the other dogs were yoked to the dogs with access to the lever, so they had no control over their experience. The dogs with control didn't seem much worse for wear; the others, however, were a different story.

When these "underdogs" were tested in another situation—where they could jump over a low fence to escape when the bell sounded—most of them didn't even try. The dogs who'd had control over the earlier shocks immediately ran away. The powerless dogs dejectedly lay down and whimpered. To the researchers, this was a completely unexpected result. They had predicted that all of the dogs would learn that the sound meant a shock was coming—and they'd all vamoose. But, in fact, these dogs had learned something much more interesting and sad: that there's nothing they can do when something bad happens, so just wait it out.[14]

Later experiments discovered that similar results can be found in many species—and that drugs that help animals to start trying again are effective antidepressants, a test that is used to this day.[15] Research

also found that animals with what became known as "learned helplessness" had similarly elevated stress hormones and lowered immune
response to those seen in Sapolsky's low-status baboons.[16] Lack of control is a key part of what makes stress damaging—and it is largely
through lack of control that chronic stress creates the health gradient
that favors those with high status.

Fortunately, however, there is an antidote even to uncontrollable
stress, one that is not necessarily out of reach for even the poorest of
the poor. And that, of course, is the kindness of others, the nurturing
contact that is designed to put the brakes on the chemistry of fear and
threat. When children are raised by stressed parents in chronically
stressful circumstances, the stress engine can become more powerful
and the brakes weaker and less able to slow it down. Nonetheless, the
capacity is almost always there; and as with all other brain systems,
patterned, repetitive experience can build it up. Practicing love is just
about the healthiest thing families and friends can do—and it's a heck
of lot easier than winning an Oscar!

twelve | WARM AS ICELAND

ASDIS OMARSDOTTIR SAT AT HER DINING room table with Maia in her cozy three-bedroom home in Kópavogur, a suburb of Reykjavík, the capital of Iceland. It was August 2008—just months before the global economic crash, which hit that country especially hard. We had become interested in Iceland because in recent years it has consistently ranked the highest—or in the top five countries in the world—on virtually all measures of health and happiness. And part of the reason for its dominance on these measures seemed to have a lot to do with low inequality, high empathy, and the way these factors allow trust to build a flourishing society. Sadly, those same qualities may have also made Iceland especially vulnerable to financial fraud, while simultaneously supporting social resilience.

Having ascertained that Maia was not an adventurous eater who would expect traditional Icelandic food like putrefied shark (it has to be putrefied; otherwise, the antifreeze-like substance in its blood that allows it to survive Arctic waters is poisonous), Asdis served a dish popular with Icelanders and foreigners alike: pasta. Asdis is an online friend of Maia's sister, Kira. They "met" because their first children were both due in the same month: March 1996. For the thirteen years since, the "March Moms"—mostly from America, but a few from

Europe—have e-mailed one another about the triumphs and challenges of child rearing.

Blond with bright blue eyes, Asdis laughs easily and enjoys quirky '80s music, knitting, and hiking in Iceland's varied volcanic, glacier-rich, and hot-spring-dotted landscape. The thirty-six-year-old mother was appalled to learn about the lack of support American families get from their communities and government when children are born. In the United States, employers are only required to offer unpaid family leave, which is not something many parents can afford to take. Many parents, out of economic necessity, return to work just a few weeks after their babies are born. By contrast, in Iceland, everyone is entitled to nine months paid parental leave for each child. Discrimination against parents who utilize this is illegal. Payment is roughly 80 percent of the person's usual salary. The time off can be split between partners. For example, a mom might take six months, while the dad takes three or vice versa. National health insurance covers everyone.

Although each American family has to cobble together its own solution to the problems of health care and childcare, most Western European countries have decided, like Iceland, that government should ensure that these services are provided to all. Providing daycare is not seen as taking sides in the debate over whether women should work outside the home—and sending children to daycare is not seen as a sign of poor "family values," or a choice to put work before family. Instead, it's viewed as a benefit for the whole society, giving every baby a good start and reducing parental stress. The Scandinavian countries see these services as best provided to all with some state involvement, not by free markets alone. As in America, the vast majority of Icelandic women work outside the home—but because of the way its social services are set up, the work/life balance issues that dominate American women's lives are far less pressing.

When Asdis gave birth to her second daughter, Sunna, the hospi-

261

tal she used had recently been redesigned to eliminate the usual bare white walls and intimidating air of medical sterility. It was now homier, with family-friendly features for new parents, like large beds in which partners can sleep together after their child is born. During the birth, Asdis was attended by a midwife. If parents wish—and Asdis chose to do this—they can stay in their private hospital bedroom for a few days after the baby is born, for free, with access to services for childcare and advice. When parents return to work, low-cost daycare in someone's home is available for children under two. The adult/child ratio is low: the maximum is one adult for four children. For children over age two, families have access to affordable preschool.

"I just think the first few months are so important," says Asdis, describing how she was stunned to learn how different it was for mothers in America and how there is no guarantee of any paid leave. "The maternity leave thing was the biggest shock for me, I [didn't] think such cruelty existed." She couldn't imagine not being able to spend this precious time with her babies—or letting political debates over whether it is right for women to work outside the home get in the way of making high-quality daycare affordable.

From its overall social welfare system to the details of its childcare provisions and hospital rooms, Iceland is a country that nurtures empathy. And, at least before the crash, Iceland consistently ranked at or near the best in the world on most indicators relating to trust, quality of life, and happiness. For example, in 2008, Iceland was the number one country on the U.N. Development Program's Human Development Index, which considers the distribution of income, health, and education levels.[1] That same year, it was also number one on the "Global Peace Index," which is calculated by comparing nations on war involvement, internal political violence, crime rates, incarceration rates, propensity for violent demonstrations, trust among citizens, and respect for human rights.[2]

Iceland's murder rate is more than five times *lower* than that of the
United States (in some years, there isn't one in the whole country).
Violent crime is rare. The country's first prison building is now the
prime minister's offices; it's main current lock-up looks more like a
hotel than a prison and was built to be a hospital. With a total population of 300,000, Iceland's prison population is only about 112, or less
than half a percent.[3] By contrast, America incarcerates some 2.3 million people, more than 1 percent of our adult population.[4] In 2006, Iceland ranked first in an international study of happiness.[5]

SO WHAT'S THEIR secret? For hundreds of years, Iceland was part of
Denmark—and known best for cod fishing or, perhaps, as a place to
see the Northern Lights. But it had long shown signs of promise: in
930, it held one of the first parliaments in the world, an early harbinger
of European democracy. The Icelandic sagas are known worldwide for
being among the first—if sometimes fantastical—recorded literary
accounts of family life and historical events. The country became a
separate state in 1919, attaining full independence during World War
II in 1944, though it was officially neutral and was first threatened by
Germany and then occupied by Great Britain.

The United States took over the defense of Iceland from the United
Kingdom in 1941, ultimately sending thousands of troops and helping
to modernize roads and other infrastructure. By 1980, the small,
highly literate country was already number two on the United Nation's
Human Development Index.[6] In the 1990s, the country dramatically
relaxed state control over its economy. Previously, all of the banks had
been government run, as were many industries, like telecommunica-
tions. While making this shift to a less-regulated free market economy, Iceland nonetheless retained the Scandinavian tradition of
maintaining a strong social safety net.

But to fully understand why Iceland was able to go from being a backwater reliant almost entirely on fishing to being such a healthy, vibrant society in mere decades, we have to look beyond these basic economic factors. A booming new area in the study of economics and sociology focuses on how trust and social structure play critical roles in determining not just individual health and happiness—but also how well regions and nations perform economically. By exploring this, we can understand more about just what is at risk when the development of empathy is threatened.

Though it might not be intuitively obvious, a great deal of a country's economic development seems to rely on how willing and able its citizens are to trust government and people who aren't friends or family. It seems completely natural to us, but every day, most Westerners casually do things that would have been unthinkable for our ancestors. We don't just take advantage of technologies like airplanes and computers that early humans literally could barely imagine—we also exchange goods and interact with strangers in ways that, even today, people from other cultures can find mind-boggling. Early humans— who spent most of their lives in small groups that rarely encountered strangers or outsiders—would probably find many of our attitudes and actions incomprehensible.

For example, to get around Iceland, Maia rented a car. Consider the amount of trust involved in that one simple transaction. You go up to a desk run by strangers, show them a piece of plastic, fill out some papers, and soon drive off in a machine worth tens of thousands of dollars. To people from some tribal societies and developing countries where outsiders and strangers are seen as untrustworthy or even as likely enemies, this seems bizarre. How can rental agents trust that someone they don't know at all will return their costly property? And why would that stranger bring it back? The people who rent the cars aren't kin or friends—why wouldn't someone take advantage of these crazy, naive foreigners?

Of course, to us, this is a perfectly mundane transaction and it is those kinds of questions that seem absurd. We know there are laws against stealing that are reliably enforced and that regulations exist to ensure auto safety. Nonetheless, we rarely think about how many strangers we almost blindly trust on an average day. When you rent a car, you trust first that the manufacturer has made a safe product—it's difficult to say how many people are involved in every step in that chain, but easily hundreds. You trust that the rental agency has appropriately maintained the vehicle. You trust that the rental agent won't steal your credit card information. And that the credit card company won't overcharge you. Before you've even driven out of the parking lot, you've placed your trust in hundreds—possibly thousands—of anonymous strangers.

Once on the road, of course, you have to trust thousands more people to obey traffic laws, pay attention to the drivers around them, and choose not to deliberately smash into you. (In Iceland, driving can be challenging for foreigners because outside Reykjavík, many roads are not paved because they are so little traveled!) And all of this doesn't even take into consideration all the structures of trust involved in things like maintaining roads, enforcing the traffic laws, and, on an even more abstract level, ensuring that money is actually worth something. Though there are obviously systems in place to try to keep everyone honest, without this kind of widespread trust, these systems would rapidly become overwhelmed and could break down. The role of trust—and the perils of dishonest and self-serving behavior by those entrusted with large sums of money—has become all too clear in both America and Iceland in the current economic crisis.

The fact that in the ordinary course of life, we rarely think about these things shows how deep the cultural assumptions we make about trust run. Trust, however, relies on human relationships, and as we've seen throughout this book, those relationships rely on empathy. Without empathy, there can be no trust, because if we don't believe that people will behave in a way that predictably honors their commitments to us, we can't connect with them or engage in mutually beneficial transactions. But to fully understand how empathy is critical to our economic well-being—and how our economy also influences our ability to empathize—we have to examine some recent research.

Over the last several decades, an enormous amount of scholarly attention has gone into understanding trust and what has been called "social capital," a term for the economic value associated with people's connections with and trust in one another and their capacity to form and utilize networks of those contacts. Overall, it can be seen as a measure of a society's ability to move beyond "us versus them" to see most people as part of "us," and therefore, as worthy of trust and as equal trading partners. It is measured by examining social networks and trust levels, as well as other factors.

However, there are at least two types of social capital. "Bonding social capital" involves relationships within one's family and tribe. "Bridging social capital" refers to connections and trust outside those boundaries.[7] It is not possible to have bridging social capital without starting with the bonding form: if you can't even trust your own family and friends, it's hardly likely you'll find outsiders any more trustworthy. Some of the most important unanswered questions in this research involve exactly how cultures can move beyond bonding social capital and into bridging social capital as well. Nonetheless, in states or countries like Iceland with the highest levels of total social capital, there tends to be much less crime, much less corruption, and higher measured levels of life satisfaction.

Societies low in social capital—or that feature only the bonding variety—on the other hand, tend to be and remain riven by tribalism, wars, desperate poverty, and underdevelopment.[8] If you can't trust people outside your own family or social group, you can't do business with them. If your transactions always aim to benefit your group at the expense of others, you wind up with corruption and graft and an inability to cooperate on the large scale needed to be competitive on the global stage or even to build reliable national institutions. A society's lack of trust essentially taxes every transaction: the more safeguards, surveillance, physical barriers, notaries, passwords, police, lawyers, courts, and regulations you need to keep people honest, the more each transaction costs and the more time it takes to get anything done. In countries with low trust, simply setting up a business or building a home can involve weeks or even months of bureaucratic wrangling and can require multiple bribes if anything is to move forward at all. Expanding the circle of empathy beyond your own family and social group is critical to a well-functioning nation.

EXPLORING THE IMPLICATIONS of these findings can help us see just how critical empathy is to how society works—as well as how a culture's foundations can be threatened when trust breaks down. Unfortunately, markets themselves can encourage behaviors that can undermine empathy and trust. They can escalate or reinforce inequality. And a market economy where people are obsessed with gaining status through wealth and possessions above all else—one that sees winning as "the only thing"—will tend to undermine trust and increase cheating.

More subtly, markets can create technologies, like television, that can distance people from one another in several ways. One obvious way is if people start to spend more time watching television than socializing; another one is through advertising that attempts to sell products by making people feel insecure about their relationships. These forces can build on one another so that, over time, people become less and less comfortable socializing. This would make them likely to watch even more television—and because they do that, they become even less connected to others and so on. By creating substitutes for social behavior like television, markets can undermine the trust they need to function.[9] Of course, social media like the Internet can counteract this in some ways by bringing together people who share interests—but these same media can also act against relational health when online socializing replaces "real life."

When factors that reduce trust go unchecked, the whole system can become vulnerable to collapse. The current economic crisis is, in part, an example of this. But to really understand why social capital matters and what it means for our future, we need to examine how trust is modulated by the brain, how that affects our interactions, and how economic inequality reduces empathy and drives greater inequality.

The brain chemistry of trust involves our old friend oxytocin. That's the chemical that helps bond parent to child, friend to friend, and lover to lover by associating the other person with comfort and safety. The fact that stress is modulated by positive social contact is not just important for understanding our friendships and family relationships—but critical to our economic relationships as well.

A series of experiments demonstrated this starkly. In one, conducted as a collaboration between researchers at the University of Zurich and the Center for Neuroeconomics at Claremont Graduate University in California, people were given an opportunity to possibly quadruple their money if they shared it with a stranger—at the risk of losing it all if the stranger chose not to return any. Twice as many of those given a spritz of oxytocin nasal spray invested the maximum sum, compared with those given placebo.[10] Another study found that oxytocin made people more generous. In this research, the first subject was given ten dollars to split with someone he or she didn't know. The second person could either accept or reject the split—but if he rejected it, neither participant would get any money. Those given oxytocin who played the role of the first subject were 80 percent more generous than those on placebo.[11] Raising empathy via oxytocin eases social connection and, therefore, trade.

In addition, other research found that being given oxytocin made people more likely to forgive someone who had betrayed their trust in this type of experiment. On placebo, people didn't give a stranger who had not returned their money in the first round a second chance, but on oxytocin, they did.[12] All in all, oxytocin appears to be the oil that keeps social machinery running smoothly, keeping us connected even when we experience inevitable friction. Obviously, these experiments suggest situations in which being too trusting can be problematic as well—but the point here is that trust involves the chemistry of bonding. We can't fully understand the underpinnings of the economy without recognizing how stress affects social life.

Oxytocin has also been linked with being trustworthy when someone has shown trust in you. In another study using the same paradigm, people who received money from a stranger and reciprocated rather than just taking the money and running had higher oxytocin levels than those who were just given an amount of money determined randomly by a computer. In other words, having higher levels of oxytocin is associated both with having experienced someone trusting you and with behaving honorably in response. This study couldn't demonstrate whether experiencing a rise in oxytocin in response to being trusted made people trustworthy or whether people who were already trustworthy were more likely to have such a response, but it did show that just being given money by a computer isn't associated with an oxytocin response.[13] To raise oxytocin, the money had to come from a real person who chose to be generous with it. Trust and at least a tiny sense of human connection had to be involved.

Further, another study did find that oxytocin levels rose in direct proportion to the generosity of the other person in the exchange: more generosity meant more oxytocin.[14] Interestingly, oxytocin has few noticeable psychoactive effects—people given the oxytocin in these studies usually weren't aware whether they'd received the drug or placebo, so it wasn't making them generally mellow. Oxytocin does, however, reliably lower heart rate and blood pressure, reducing the activity of stress hormones like cortisol. So, here again, we see that oxytocin is the chemical through which social contact relieves stress. It makes us comfortable with one another, open to both relationships and exchange.

To anyone who isn't an economist, this research makes sense. We're social creatures and if someone who seems trustworthy offers us free money to share, we'll try to be fair. We expect others like us to be that way as well—this allows our world to work. Many people seem to be built this way, across cultures, though there is variation, of course, related to who is allowed into the circle of "us" and who remains one of "them." Traditional economics, however, wrongly predicts the outcomes of these experiments. It sees human nature as essentially selfish, seeking above all to maximize wealth, even at the expense of relationships. From this perspective, there's little to be gained by trusting a stranger who probably sees other people in the same way: as objects of exploitation.

Curiously, this unempathetic view of the world is reinforced by imagining (via the perspective-taking aspect of empathy) that everyone else is similarly selfish! Traditional economics also predicts that if you give two subjects ten dollars to share, contingent on the second party accepting the split chosen by the first, any sum at all should be considered acceptable. Some free money should always be better than none. Instead, in real life, if the first person decides to split ten dollars by keeping nine dollars and offering just a dollar to the second, that isn't what happens. Most people will refuse the one dollar to ensure that the greedy stranger doesn't get any money, either. A low offer is seen as an insulting lack of reciprocity and experienced as unfair.

Our brains have evolved, it seems, not only to see the world through multiple perspectives, but also with a bias toward fairness and relational connection. This is why most children immediately get the concept of "that's not fair" without ever being taught to measure whether their sisters are getting bigger scoops of ice cream. Even vampire bats will exclude fellow bats who don't share—and even economics students don't respond the way their professors' theories would predict (though, interestingly, they are greedier than those who choose to study other subjects).[15]

SO HOW DO countries get—and maintain—the trust and social capital needed to function well? As we saw earlier, empathy grows first in the family and is learned starting with the parent/child relationship, moving out to other relatives, friends, and from there into the larger social world. Asdis's ability to spend the time she wanted with her babies in their first nine months of life without worrying about her job surely reduced her stress levels—and this simultaneously made it easier for her to healthily modulate her children's stresses. That alone could help create an empathetic culture. What happens in the family doesn't stay there—it can affect the tone of whole societies. Mostly, of course, families do fine and children grow up to trust and interact as their parents did.

But sometimes, natural disasters, wars, starvation, and other major traumatic events can leave their mark on a whole generation, damaging its ability to trust and to trade and leading to eras with little social or economic growth, as we saw. And, of course, the most severely abusive and damaging families can sometimes produce children whose lack of empathy is a danger to everyone else—or children who, unlike in Trinity's story, are not able to overcome and function. We can trace this right down to oxytocin: more early nurture means more oxytocin, higher stress means less. Over a longer time scale, decreased opportunities for social connection and learning can also wear away at the foundations of empathy.

Iceland, in contrast to the countries where cycles of trauma create mistrust, was in the opposite situation. During the early twenty-first century, it had a surfeit of social capital. As a northern island nation in which survival has always been quite visibly dependent on cooperation, it had already built a deeply socially connected society. There was not much inequality—and people valued the idea that everyone, not just a few, should benefit from increased economic growth. Says Mary Gordon, describing growing up in Newfoundland, also a northern island, "Quite honestly, because it was an island, people lived with great vulnerability and levels of interdependence. You're hanging on to a cliff for survival in those fishing ports. There was a very deep degree of commitment and connection. And whether you did it out of self-interest or because [other people] might be needing it, I don't know." Geographic isolation and the dependence on others that comes from having to survive harsh conditions together can increase social cohesion.

This underlying sense that "we're all in it together" means that people like Asdis share not only an excellent social safety net, but also form strong family connections. Asdis is the second of five children—she has three sisters and a brother. "I try to meet every one of them every week; I'm very close with my siblings," she says. How many Americans even talk on the phone that often with their immediate family members? But this level of connection is common in Iceland.

Asdis's parents both work in research labs, her mother studying cancer cells and her father doing veterinary research. Iceland also has excellent literacy and education rates, and most of the population is fluent in both Icelandic and English, which is taught in middle school. When Maia visited, Asdis was just finishing exams at a junior college where she studied sociology. She graduated in 2009.

Being raised around so many close relatives, in a culture that values education and takes care of those who need extra help, is an important part of generating social capital. Iceland has another advantage here as well: it is extremely homogeneous, with very little racial, ethnic, or religious diversity. Though diversity can be a positive factor in many other ways—and is something a globalized world has no option but to confront—in terms of ease of empathy, it can sometimes be a barrier. In Iceland, 93 percent of the population is Icelandic: there are no significant minority groups who might be targeted as "others" or "outsiders" and become a focus for us-versus-them thinking.

Though multicultural societies have learned ways of getting around this barrier, because of our evolutionary roots in small, homogeneous groups, it is easier to create empathy among people who look alike, who visibly seem "like us." We might unconsciously see them as relatives. Indeed, an average Icelander walking down the street in Reykjavík is likely to be genetically related at least distantly to virtually all of his countrymen around him. Iceland's people are so interrelated that several important genes have been discovered by studying its population, which is easier to do because of the reduced variation.

It makes sense that people who look similar would be more empathetic toward one another: kindness toward kin is the first kind of kindness, and people who look like you are actually more likely to be related to you. Research confirms that we are more likely to empathize with people who look like us.[16] Given empathy's roots in mirror neurons, a more similar "reflection" is probably easier to imitate and resonate with. This evolutionary bias toward visible and genetic similarity can also help explain why Icelanders would have little objection to spending on things like universal health care and family leave:

when people imagine benefits going to their families and people like them, they are much more likely to support them. Conversely, when considering spending taxpayer dollars on foreigners or people who are seen as "outsiders," there tends to be far less agreement, especially when money is tight. So Iceland's unity in terms of ethnicity, religion, and culture helps create and sustain social capital.

Of course, this aspect of social capital is not one that multicultural societies like America could or would even wish to replicate. We're racially and ethnically and religiously diverse; we also have many regional and linguistic variations. Despite these differences, we have historically been a country with high levels of trust of outsiders that has enabled massive economic growth. So why have we been able to create this kind of multicultural social capital, when other countries have had much more difficulty with the trust needed for trade? This question is one of the most intensely studied and debated areas of economics and development. A huge variety of factors from geography to religious values to historical vagaries of government development and structure are seen as having an influence.

Both a sense of similarity and at least the appearance of equality of opportunity, however, seem to be essential. One very important way of creating this sense that "we're in this together" is to keep differences in wealth from becoming too extreme. People who share socioeconomic status tend to have a lot in common. So the smaller the distance a society keeps between rich and poor, the greater the likelihood that its people will be able to connect with one another.

Historians and economists have long known, of course, that great differences in wealth produce unstable societies. When wealth is extremely concentrated in the hands of a few and the rest live in humiliating poverty, it's pretty obvious that the rest of society isn't likely to thrive and that political unrest is a serious risk, in the absence of repressive measures. The peasants with pitchforks will eventually threaten the manor. But what researchers have only recently begun to discover is how this affects physical and mental well-being. It turns out that extreme economic inequality is not only bad for the poor, but for the middle class and wealthy as well.

Iceland, you probably won't be surprised to learn, has typically been extremely low in economic inequality. The Scandinavian countries, in fact, repeatedly top the world charts for having the lowest rates of economic inequality. That they also consistently rank high in health and happiness is probably not coincidental.

But how would low inequality contribute to social capital? The effects on empathy of having a highly interrelated population where everyone looks similar are just the first part of the answer. To understand the second part, we have to return to America for a short expedition. The profound effects of inequality on social capital can be seen through the lens of the story of Wendell Potter, a former high-level health insurance executive. He recently—via an experience of empathy—became a strong advocate against the industry in which he spent much of his career and in favor of reforming the system to provide health care to all Americans.

WENDELL POTTER WAS born and raised in Appalachia, in East Tennessee. He attended the University of Tennessee in Knoxville, studying communications. After an early career in journalism, he switched to public relations. He spent fifteen years at Cigna, one of the top five largest health insurers in the United States, becoming its head of corporate public relations and chief public spokesperson. Potter is a businessman: he believes in the American system of capitalism. Although he recognized intellectually that our health-care system does leave some people out, for most of the time he worked at Cigna, he felt that good insurers could ethically profit by improving health care and cutting unnecessary costs.

In the mid-2000s, however, he began to have qualms about the morality of his work. During the early days of managed care, it had been relatively easy for companies to find ineffective, expensive services that could be cut back and not only save money for the insurer, but actually improve the health of patients. Then, however, Potter began to see increased emphasis on profit—with less concern for patients. In fact, insurers started shifting away from managed care and toward high-deductible products that did little more than make patients pay more for less. And he was out there selling this to the media and the public as being good for them, far better than any system controlled by "government bureaucrats." He was, in fact, fighting intensely against any legislation that might cut insurance-company profit, even if it would benefit patients. He helped defeat the Clinton administration's health reform efforts. Though he felt some unease, he was still able to reassure himself that what he was doing was acceptable.

But one day, on a visit to his parents and his hometown, he decided to check out a local health-care "expedition" in Wise, Virginia. It was held in a muddy field on the county fairgrounds, not far from where Potter grew up. Potter expected booths with nurses checking for high blood pressure, nutritionists giving diet tips, and well-meaning pamphlets on exercise and quitting smoking. What he didn't imagine was a scene that looked like it belonged in a third-world country, with desperate people showing up from numerous states and waiting in seemingly endless lines in the rain for basic health care. He didn't realize that the strange name of the program came from the fact that the "expedition" was sponsored by a group that originally did foreign aid. Founded by the former host of TV's *Wild Kingdom,* its traditional "expeditions" helped people in developing countries whose access to health care was limited not by insurance companies, but by subsistence-level poverty.

"What I saw were doctors who were set up to provide care in animal stalls," Potter told PBS's Bill Moyers in a TV interview, still incredulous. "Or they'd erected tents, to care for people. I mean, there was no privacy. . . . I've got some pictures of people being treated on gurneys, on rain-soaked pavement. . . . People drove from South Carolina and Georgia and Kentucky, Tennessee—all over the region," he said.[17]

If he'd been anywhere else in the country, he might have dismissed it as a sad effect of entrenched poverty, nothing to do with him or his business. But this was literally close to home—the guy on the gurney could have been a friend from high school, for all he knew. "There could have been and probably were people that I had grown up with," he said. "They could have been people who grew up at the house down the road. . . . And that made it real to me."[18]

Potter says he felt like he'd been hit by lightning. "It was almost— what country am I in? It just didn't seem to be a possibility that I was in the United States," he told Moyers. In his job, high above Philadelphia in a cool, climate-controlled office, he was never exposed to the impact of his company's decisions. He'd literally been insulated from them. People of his executive class never visited muggy, rainy county fairgrounds or even regular doctors' offices for health care. They had "concierge services"; they never waited for anything, even in a plush waiting room. Behind the dark glass in air-conditioned chauffeured cars, they didn't have to see at what costs their profits had been taken.

Several weeks later, Potter was sitting on a corporate jet, in a spacious leather seat being served by a flight attendant. His food was presented on a gold-rimmed plate. He was handed gold-plated silverware—no plastic antiterrorist sporks there! "And then I remembered the people that I had seen in Wise County. Undoubtedly, they had no idea that this went on, at the corporate levels of health insurance companies," he said and felt ashamed.[19]

This is how inequality reduces empathy—it places both physical and emotional distance between the rich and the poor. In America, we tend to think more about racial prejudices as barriers, and of course, these do matter tremendously as well. But people often discover that they actually have more in common with someone of a different race of the same socioeconomic status than they do with someone of the same race but in a different class. Although virtually all Americans consider themselves middle class, it is not hard for most of us to instantly guess a great deal about someone else's wealth simply by paying attention to dress, appearance, and language use. And cross-class interactions have become increasingly rare, as the wealthy live in different neighborhoods, attend different schools, and move in increasingly isolated social worlds, like Potter.

The differences have become so distinct that they can almost be seen as different cultures, where it is hard to know how to interact because members of each group don't know the unspoken rules that guide the behavior of the others. Empathizing across this gap and not taking an "us" and "them" view can become extremely difficult. An unexpected interaction with someone from another class—outside a work context where the roles to be played and people's expectations of one another are understood—can raise our levels of stress hormones because they seem so different. Neither the richer nor the poorer person knows exactly how to respond to the other. It was only the fact that Potter visited the health-care expedition near his hometown that broke through this barrier for him: he empathized because he saw folks from home as "us," not some alien and undeserving poor "them."

In the United States, this situation has been made worse by our own desire not to see class as an important factor. Though we like to think that class distinctions don't matter here, in fact, research shows that in the last few decades, a child born poor in Europe or Canada has had a greater chance of becoming wealthy than one born poor in the United States.[20] Think about this. If we aren't providing greater opportunities for all to advance, how can we justify offering only the barest safety net to those who don't "better themselves"? These days, poor people actually have a better chance of striking it rich in what many Americans see as stultified "old world" Europe and "boring" Canada. As Potter's story shows, the wealthy here now live in a completely different world.

What's more, the American dream that anyone can rise to the top with sufficient hard work and drive is probably a key part of what has traditionally allowed us to trust one another and be an "us," not a bunch of "thems." That dream gives us the sense that we all have the opportunity—if we work hard—to do well here. This has allowed us to sustain a sense of solidarity and common identity, even though we could always also see that we really aren't a classless society. Despite our focus on rugged individualism, historically Americans have been rugged individualists together, sharing this value, as de Tocqueville and many others have explored.

So how did our dream start to die? An important part of the answer is the speed and intensity with which economic inequality has recently grown. As of 2008, the richest 10 percent of the American population held 71 percent of Americans' net worth.[21] But this was not always the case. Before the Civil War (excluding, of course, inequities related to slavery and women's social roles), the United States had only a small wealth gap between rich and poor. The Gilded Age (1870s to early 1900s) of the "robber barons" brought massive inequality—but that ended with the Great Depression and New Deal policies that helped level the playing field. By 1942, the top 10 percent of Americans earned about one-third of national income. The Reagan years and go-go 1980s saw the start of a new rise in inequality. By 2002, the very rich were receiving 50 percent of American income.[22] The current chasm between rich and poor is the largest ever measured in the United States—and the last few decades have seen dramatic increases. The advantages of being born rich have grown—at the same time as the overall proportion of wealth held by the rich has increased. Meanwhile, the middle class and the number of well-paying jobs is shrinking. The 2008 crash probably reduced inequality some by stripping away a lot of excess wealth from the rich—but that has hardly improved conditions for everyone else. Concluded former banker and author Charles R. Morris in a prescient 2006 *New York Times* op-ed: "An economy has psychological or, if you will, spiritual dimensions. A conviction of fairness, a feeling of not totally being on one's own, a sense of reasonable stability are all essential components of good economic performance."[23]

Understanding empathy and its role in social capital is the key to recognizing how corrosive a lack of these qualities can be. Because, as it turns out, economic inequality is inversely correlated with trust. In other words, the greater the gap between rich and poor, the lower the level of overall trust there is in a society. A large gap between rich and poor typically characterizes nations that stagnate and do not see large improvements in people's health and happiness over time. For example, these are some of the nations with the world's highest levels of inequality: Iraq, Afghanistan, Congo, Sudan, and Chad. Many of the world's "trouble spots" are marked by the largest gaps between rich and poor.

And in recent years, the United States has seen a rise in inequality that is moving it out of the company of relatively equal European countries and Canada—and toward that of highly stratified places like Mexico. As in Potter's story, there are increasingly fewer places in America where rich, poor, and middle class interact spontaneously, as equals—and few situations in which unplanned, noncommercial friendly interactions take place regularly. Public schools—which used to be a great equalizer—are becoming more and more economically segregated each year. This is related to increasing geographical segregation between the middle class and the poor—and all of this usually includes racial segregation as well. Two-thirds of American children have family incomes that place them in the middle class—but 25 percent are educated in schools where the majority of students are poor.[24]

Access to technology and the decline of mass media have also increased class distance. When everyone no longer watches the same network news and entertainment shows, when only 40 percent even subscribe to premium cable and satellite TV,[25] and when about 25 percent of the population doesn't have access to the Internet at home, even the formerly shared language of popular culture breaks down. This high level of disconnectedness is exactly the opposite of the situation needed to create empathy and sustain trust.

The greater the differences between people, the less familiarity they have with one another, the fewer cultural references they share, the less likely they are to trust one another. Economic inequality increases this distance in multiple ways. Probably the most destructive of these is the cynicism that develops from believing that the system is rigged in favor of the wealthy. This automatically sets up an us-versus-them dynamic. If the system is unfair, why not try to cheat it? If the rich are "getting theirs" by cheating the rest of us, why shouldn't the poor fight back by slacking off or by outright theft? From the perspective of the rich, if the poor are lazy and criminal, why should we help them? Once people begin to see the system as protecting the already-rich, and both sides start to believe that the other isn't playing fair, the trust that is necessary for them to engage with each other breaks down and a vicious cycle of decreasing trust and increasing corruption begins. This situation is one that many developing countries have been unable to escape.

High inequality, not surprisingly then, is correlated with high crime rates. For example, the murder rate in Iceland in 2004 was 1 per 100,000 people, while the U.S. rate that same year was more than five times greater: 5.5 per 100,000.[26] Iceland has 39 prisoners per 100,000 people; the United States has more than eighteen times that, with 726.[27] The effects of inequality on crime can even be seen on the state level within the United States: a study looking at the year 1990 found high variability of murder rates within America, with some states having rates as low as 2 per 100,000, while others were as high as 18 per 100,000. The biggest single factor influencing these rates was income inequality.[28]

High inequality is especially closely linked with corruption—the countries with the least inequality also tend to have the least corruption. Four of the countries that have the lowest levels of corruption on Transparency International's Corruption Perception Index for 2008 are Scandinavian, for example, with Iceland ranked at number 7.[29] The United States is number 18—while the bottom ten countries include Somalia, Afghanistan, Iraq, Sudan, Haiti, and the Democratic Republic of Congo. All of these countries are extremely impoverished and have had lasting difficulties in terms of developing a working economy.

Inequality doesn't just affect crime and warfare, however. It also profoundly influences both mental and physical health. Both life expectancy and infant mortality rates are negatively correlated with inequality: the greater the inequality, the lower the life expectancy and the higher the infant death rate. Iceland, for example, has one of the highest life expectancy rates in the world—80.5, compared with 78.14 for the United States. Iceland comes in at number 14 in international rankings on life expectancy—while the United States is a distant 45.[30] In terms of infant mortality, the United States loses 6.26 infants per 1,000 live births—while Iceland, which has the lowest rate in the world, loses just 3.23 per 1,000.[31]

But how could something as abstract as economic inequality affect something as concrete as infant death rates and overall life expectancy? Once again, it comes down to the social aspects of our stress systems and, particularly, how they respond to the power differentials explored in the last chapter. You might think that this would be a more straightforward relationship, connected simply to reduced access to health care among the poor in stratified societies. For example, Asdis's access to free, high-quality medical care and childcare would give her an obvious health advantage over an American mom with gaps in health insurance coverage who struggles to afford basic coverage and childcare. However, it turns out to be much more complicated because studies have found that while access to affordable services helps, this doesn't fully account for the differences between countries with high and low inequality.

Instead, these differences tend to track imbalances in social status and the ability one has to control the circumstances of one's life. Simply having higher income would seem to be the most important factor, because wealth should provide access to more advanced technologies and better medical care. But inequalities remain important, even for countries with a high per capita income. In such countries, there's no connection between average per capita income for the country and life expectancy and infant mortality. Both of these rates are linked to inequality, however, as greater inequality increases infant death rates and shortens overall life span. Even among the rich, there is shorter life expectancy in high inequality countries than there is in more egalitarian places.

The reasons for this bring us back to the human stress system and the way it is affected by hierarchies and control. As we saw in the last chapter, most of the research in this area has found that people lower in socioeconomic status have death rates around two to four times higher than those of higher status, no matter what cause of mortality is examined.[32] Although higher rates of smoking and other unhealthy behaviors among the poor account for some of the difference, even when these are factored out, large disparities remain. These disparities have been found in all countries studied, although they are decreased in more egalitarian nations, like Iceland. They result from how status struggles affect the stress response system, and how chronic, uncontrollable stress damages the brain and body. And this can be seen in Iceland, even in the way it diverges from conditions that normally would seem likely to be linked with health and happiness.

Even before the economic crash in Iceland, there was one anomaly in the otherwise rosy picture of this small, strange country: Iceland has a high divorce rate. Although its divorce rate is about the same as ours, compared with many other European countries, Iceland's rate is high. It is 37 percent.[33] Moreover, 65 percent of births in Iceland occur out of wedlock, and 15 percent of all families with children are headed by a single parent.[34] So there are many parental separations that occur without being recorded in the official divorce rate.

Asdis's family reflects this trend. Her oldest child, a daughter named Olof, was the result of a relationship she had before she married her current husband. Asdis met Olof's father in an online game, which, it turned out, they were both accessing from the same university computer center. They fell in love quickly. The couple moved in together within two weeks, living in an apartment owned by his family, who were staying in Sweden at that time. Asdis became pregnant and Olof was born when her mom was just twenty-three. Icelandic women tend to start having children much younger than women in the rest of Europe and the United States, and it may not be coincidental that Iceland is one of the few European countries to have a fertility rate above the replacement level. "I think the maternal instinct is quite strong in us," says Asdis, explaining why she thinks Icelandic women prefer early maternity.

They are also less concerned about the economic and career

impact of child rearing. In Iceland, 80 percent of women work outside the home—one of the highest rates in Europe—and parental leave and daycare access ease work/life balance stresses. Icelandic women don't have to worry about establishing themselves first or getting derailed onto the "mommy track" at work: they can have children when it is biologically easiest and not put either their ability to spend time with their babies or their chances for success at work at risk. High-quality daycare like Iceland provides is not linked with pathology— children attending have better vocabularies than those who don't. One major study found a slightly higher chance of misbehavior at school in children attending daycare compared to those staying at home, however this was within the normal range.[35]

But when Olof was still a toddler, Asdis and her partner broke up. The split was amicable, and Olof spends every other weekend at her father's house; she can also call or see him at other times if either wants to do so. Olof's dad has two other daughters by two different women, one of whom is now his wife—and the two daughters who don't live with him visit on the same weekend, with little friction between the parents. Says Asdis, "For birthday parties, he comes over with his wife, and I more often see his kids' parents actually than I see his kids. I get to see them regularly. They always seem really nice to us."

SO HOW COULD Iceland nonetheless remain such a happy country? Shouldn't all that splitting up and reconfiguring of families harm children? Doesn't having children by multiple fathers put little ones at risk from abuse by stepparents? It turns out that in a country with high social capital—including, particularly, strong extended families like Asdis's family—the answer is not quite what you would expect. The effects of family disruption seem to be buffered. Icelanders don't seem to be as wounded by the breakdown of nuclear families as Americans are.

One part of this buffer is the social safety net of public health care and daycare—access to which is not affected by a person's marital status and current employment. But another, possibly more important factor is the fact that if you have a large, strong social network, the loss of any one link in it is simply not as devastating as it is if your network is smaller and weaker. If, as Sarah Hrdy's research suggests, humans evolved as cooperative breeders, a strong extended family might ultimately be more important to a child's health than an intact nuclear family, though obviously having both is better.

In contrast, America puts a tremendous amount of weight on its nuclear families and the marriages at their center. If a young child loses a parent to divorce in this context, the harm is magnified exponentially. Here, there are often no nearby extended family and friends to help everyone pick up the pieces. In contrast, Iceland is a very small country. If you have not only your mom and dad in your life in a daily or weekly context, but also your grandma, aunties, uncles, grandpa, mom's friends, and cousins, the impact of reduced contact with your father is greatly mitigated. Research suggests that children of single parents with strong extended families and other social support do better;[36] in our past, as Hrdy's work suggests, this may have been even more important to children's survival than the presence of a father.

Icelandic children are, of course, not unaffected by their parents' relationships and breakups. Like anyone else, they can suffer profoundly from family dysfunction. However, in countries like Iceland, the impact of divorces and breakups is much smaller because the child's life afterward tends to continue to involve both parents and both extended families, particularly on the maternal side. As we saw in Chapter 8, child rearing has long required more than just a mom and dad; the more high-quality, consistent relationships in a child's life, the more resilient a child will be to any kind of loss. This doesn't mean that the loss of fathers is good, of course, nor does it mean that they don't play critical roles in child development. But the conditions that create empathy and trust reinforce one another—and as long as social networks remain strong, people are resilient.

The long-term effects of Iceland's economic meltdown, however, remain to be seen. In 2008, Iceland's currency, the krona, lost virtually all of its value. Unemployment skyrocketed, rising from 1.5 percent to 8 percent in months.[37] Both Asdis and her husband lost their jobs. Many people who had taken out loans to buy expensive cars like Range Rovers—now worth far less than the balance due—actually burned their cars to try to get insurance payments. Behind the scenes, it turned out, Iceland's egalitarian premises had been violated—a small group of bankers and financiers had indeed become much richer than everyone else. Part of the reason that they were able to get away with their risky moves for as long as they did may have been because of the high general trust that the country had built up.

Now the effects of their gambling had bankrupted Iceland. Large crowds gathered in freezing weather in downtown Reykjavík, infuriated by the government's failure to protect the economy. Shortages of essential items were reported in some stores. For days, the demonstrations continued. There was clear anger, even fury. But, interestingly, there was no violence. Demonstrators shouted and even threw food at government buildings—but that was about as disorderly as it got. They soon voted out the officials who had gotten them into the mess.

For Asdis's family, at least, the crash wasn't all bad. Their experience was marked by an act of compassion. Unprompted, a wealthy Dutch friend—whom Asdis knew only online, through gaming—contacted her when he heard about Iceland's situation. He sent 100 euros and an ATM card, saying she should use it if her family ran into trouble. "Thankfully, we haven't had to use his kindness but it's there if we need it," she says. As of November 2009, both Asdis and her husband had found new full-time jobs.

She adds, "I think most people here are all right, actually. Last October, we completely changed our lifestyle because we didn't know what was going to happen. For first time in our lives, we now have a lot of money in bank. We've been able to save up." The crash has also allowed her to spend even more time with her family—something that she says has made her happier. Sunna and Olof are not particularly thrilled that their parents have cut spending on their clothing and toys—but while the economic crisis originally frightened them, they have adjusted. While unemployment in Iceland was 7.6 percent as of October 2009, the American rate was significantly worse: 10.2 that same month.[38]

Whether their strong social capital can sustain Iceland's social services and high standard of living—not to mention world happiness and development records—is an open question, but it means they are far better equipped to be resilient than most.

thirteen | ALL TOGETHER NOW

SO HOW DID WE GET FROM Mary and baby Sophia to bread-throwing unemployment protests in Iceland? What can we really learn from Eugenia's experience in a Russian orphanage that can help us understand the darkness of medieval Europe? How do Danny's family of con artists and liars and Brandon's early life in front of a TV set and all the other stories told here fit together—and what can this tell us about parenting and policy?

The unifying thread is our fundamental yet developmentally vulnerable capacity for empathy. Empathy—fully expressed in a community of nurturing interdependent people—promotes health, creativity, intelligence, and productivity. In contrast, apathy and lack of empathy contribute to individual and societal dysfunction, inhumane ideologies, and often brutal actions. All of the stories in this book show how experience profoundly affects the development and even the most basic ability to feel and express empathy. Some show explicitly how certain experiences can be passed down—through both genes and the parental environments they create for the young—from generation to generation. Life in the womb and the cradle is intimately shaped by the culture around the parents; parenting styles and environmental stress in turn help create both culture and children's experience of it.

Consequently, while we are born for love, we need to receive it in certain, specific ways early in life to benefit most from its mercy. We need to practice love as we grow through different social experiences to best be able to give it back in abundance. The brain becomes what it does most frequently. It is shaped every day by what we do—and what we *don't* do. If we don't practice empathy, we can't become more empathetic. If we don't interact with people, we can't improve our connections to them. If we don't ease one another's stress through caring contact, we will all be increasingly distressed. Empathy simply does not develop equally in all situations. Threats to its development are, unfortunately, common. And stress relief efforts in the absence of social support are ultimately doomed to failure.

In Jeremy's case, his mother Angela's implicit teaching first went astray with too much eye contact and too little independent experience of stress because of her fears that his facial deformity would lead to too many painful interactions; her well-intentioned efforts to spare him distress robbed him of the experiences necessary to build an independent capacity to modulate stress. In Eugenia's early life in the orphanage, there was, conversely, too much stress and not enough individual face-to-face and skin-to-skin connection. In Sam and his son Joshua's story, we saw genetic influences, sensory problems, and other issues related to autism spectrum conditions that interfered with the perspective-taking part of empathy. And while Danny and Ryan both gained mastery of the cognitive aspect of empathy, bad parenting and, in Ryan's case, repeatedly broken attachments stunted their development of empathy's warm emotional core. In Trinity's triumph over a childhood of trauma and stress, however, we also saw that empathy can be resilient—that it can sometimes develop and find other sources of social soothing, despite the most relationally toxic early environments.

Nonetheless, our brains evolved to expect certain social arrangements and early experiences to shape their developmental trajectories. The brain shares this reliance on particular experiences for proper growth with the rest of the body. For example, a fetus would not develop properly if its mother spent her pregnancy in a zero-gravity spaceship: the cells of our body evolved in the earth's gravity and they need its pull in order to grow in the right direction and position. Similarly, babies' brains expect to be surrounded—as they were for thousands of generations—by a multigenerational, multi-family group. Our species' unique behavioral flexibility and our ability to utilize information and memory in ways that animals cannot has allowed us to create environments—like spaceships or even plain old cities—for which our brains are largely unprepared.

Because we can pass on not just genes but knowledge and technology, we have the capacity to culturally "evolve" rapidly by reshaping our own social and physical environment. This has allowed our species to invent countless, varied ways of living together. Over the thousands of years of human history, many languages have been created and spoken; many styles of governance and sets of values and belief systems have been tried and utilized. Some of these lifestyles are very respectful of our biological gifts: in these cases, the cultural beliefs, child-rearing and educational practices, government policies and laws reflect fundamental insights about the nature of humanity. Other societies and cultures have not taken key aspects of human nature into account. Totalitarian governments—to take the most extreme examples—have failed to recognize the human need for freedom, our fundamental reliance on family and friends, and the tendency for those in power to abuse it in the absence of any checks and balances. This has caused untold atrocities, pain, and deaths. You can probably think of many other situations in which people's basic social needs have been ignored by a policy or practice and great suffering has resulted.

When a society respects human biology, its full members—at the very least, those considered "us" rather than "them"—tend to prosper and flourish. But the further we invent ourselves into worlds that attempt to fight or disregard our own biology, the greater the probability there is for extreme stress, distress, and dysfunction to spread. Unfortunately, crude biological arguments have been used in the past to justify oppression—via racist, sexist, and other claims that it is "natural" for some groups or types of people to be subordinate to others. We want to be clear that we are not talking about that here: when we conceptualize a society as being biologically respectful, we mean that the fundamental needs for social contact and interdependence of both children and adults are being met and reasonably well addressed, rather than ignored. We don't mean that "traditional" or "natural" is always good.

Once you understand that we are ineluctably interdependent, a tremendous number of seemingly disparate facts start to make sense. Why would giving to others help the giver—reducing chronic pain and depression,[1] even extending life?[2] Because kind social contact relieves distress, reducing the toll chronically high levels of stress hormones take on the body. How could self-centeredness decrease survival rates from heart disease?[3] By cutting social contact and elevating levels of those same hormones. Why is cuteness so appealing? Because it's one way that the brain's social networks link nurture and pleasure. Why do money and possessions so rarely bring the happiness we expect? Because they often distance us from one another, rather than bringing us closer, emphasizing status gaps, not narrowing them. And, finally, what causes much of life's most agonizing pain? This, too, is related to relationships—those we lose, fail to maintain, or that become one-sided or abusive. All of these phenomena are connected by the fact that brain development is utterly reliant on empathetic nurture—and that humans evolved as profoundly social creatures.

A central argument of this book is that expressed empathy has been eroding due to the rapid changes in our society that have become measurable over the last five decades. Of course, we aren't saying that the 1950s or early 1960s were a "golden age"—simply that we can measure a decline in many forms of social connectedness that has taken place since then in addition to welcome changes in civil rights for women and minorities. We believe that concurrent advances in technology, the high mobility of our populations, ongoing instability of families and communities, and compartmentalization of educational, work, and living environments have contributed to a reduction in the number and quality of human interactions below that which is necessary for the full development of our capacity for compassion.

The extent of the collapse of "relational wealth"—and the resulting rise in what we call "relational poverty"—is stunning. Despite the long lists of Facebook friends that many people display, multiple lines of evidence suggest that Americans overall are now less trusting, have fewer close relationships, spend less time socializing, and tend to keep their children in structured activities that reduce the time they spend with friends. We give a smaller percentage of national income to charity, participate less in most political activities, and join fewer clubs or groups that require face-to-face participation. Religious participation has declined as well.[4] All in all, we have fewer daily opportunities to connect face-to-face in shorter amounts of time with smaller numbers of people.

Consider again that the percentage of Americans who say that "most people can be trusted" dropped from 58 percent in 1960 to 32 percent in 2008.[5] That means that the majority of Americans now view most of their fellow citizens with suspicion, not as potential friends or reliable business partners. Not only is this a cause for concern in terms of social life—but, as we saw in the previous chapter, the level of trust a country has in itself and its government is strongly related to its capacity for economic growth. A recent study found that trust in government—more than any other factor—explained the variation of murder rates over time in the United States and Western Europe.[6]

In America, regional differences and intensely partisan political behavior have also increased, enhancing the "us-versus-them" part of politics while decreasing cooperation. Shrill rhetoric from both parties has made bipartisanship challenging: how can you ethically collaborate with people whom you have labeled as unpatriotic traitors set on destroying the country?

Families are shrinking, too. Right now, about 33 percent of American families are single-parent families[7]—and a much larger group of American children will spend at least some time living in a single-parent home. Not only are there more single-parent families, but even two-parent families are getting smaller, with greater numbers of children growing up as "onlies." Of course, single parenting and only children aren't themselves necessarily problematic—but by basic arithmetic, having more people in the home means more opportunities for social interaction. Being extremely close to several different people increases your opportunity to learn about human quirks and variations; having more than one parent or sibling provides repeated chances to learn to empathize with other kinds of people. If these interactions are negative—or if there are too many children for parents to adequately nurture—they may be harmful: we are not saying that staying in abusive or conflict-filled relationships is good for children or that single parenthood is, of itself, always detrimental. But fewer parents and children in the home often means fewer relationships. Fortunately, as we've seen, having strong networks of extended family and friends can make children from both single and nuclear families more resilient.

Unfortunately, these networks, too, may be fraying. Precise statistics on people's connectedness with extended family are hard to find. Interestingly, unlike other aspects of social life, visits with relatives do not seem to have declined between the 1950s and the 1980s—and longer life expectancies do mean that children have more years in which to get to know their grandparents and even great-grandparents.[8] However, America is still quite a mobile society: 12 percent of the population moved between 2007 and 2008 alone[9] and nearly two-thirds of adults live in a community other than the one in which they were raised.[10] It's safe to say that many families—like Eugenia's—live farther from relatives like aunts, uncles, cousins, and grandparents than allows the close, frequent contact that they prefer.

Our living arrangements have contracted, too, with average household size going from six in 1850 to three or less today, as noted earlier.[11] As American houses have grown larger and larger, the number of occupants has shrunk—leaving us more and more isolated physically as well as in terms of relationship numbers. Larger families aren't being replaced by increased friendships, either. Indeed, the number of what researchers call people's "confidants" has steeply declined. These are basically close friends or family members that you trust enough to share your darkest secrets with or to ask for help in the middle of the night.

In 1985, most people said they had three such intimates—but by 2004, this number was down to two. A full quarter of Americans have no one at all to confide in.[12] And 80 percent said they only shared important emotional information with their spouse or with one particular family member—meaning, essentially, that they had no close friends outside the family.[13] This is concerning because, although divorce rates have fallen from their historic high of 48 percent in the

1970s,[14] 35 percent of first marriages still end in divorce before their fifteenth anniversary.[15] Time spent dining with or just hanging out with friends is also on the decline.

Some of the most dramatic changes here specifically affect children and childhood. The amount of time spent playing freely fell by nearly one-third between 1981 and 2003.[16] Outdoor play—which is most likely to be social—was hit even harder. The number of hours that children spend playing outside in unstructured activities was cut in half between 1981 and 1997.[17] In schools, overall time available for social activity has been reduced by cuts in recess time, in gym classes, and by shortened lunch periods. Only 57 percent of elementary school districts currently require recess for all grades, one through five—and only 13.7 percent of elementary school children have gym classes at least three days a week.[18]

Though good numbers are not available to track these trends over time—and there has been a backlash with parents recently advocating for more recess and gym to fight obesity—parents clearly recognize that today's children have dramatically less freedom to roam and unstructured time compared to their own childhoods. The concept of a "play date" was virtually unknown to parents in the '70s and '80s—but fears about safety have led today's parents to schedule children's time together, rather than having them find friends in the neighborhood. Safety fears also have produced kindergartens and preschools where physical affection toward children—which, as we've seen, is important for their health and growth—is actually banned.

Anxieties about crime and safety have also had another untoward and isolating effect, particularly for minorities: mass incarceration. As of 2008, 2.3 million Americans were in jail or prison. The "land of the free" locks up more of its own people than any other country—including communist dictatorships like China. About a third of young black men are under criminal justice supervision—incarcerated, on probation, or parole—at any given time.[19] Nearly two million children, a quarter of whom are age four or younger, have parents in prison.[20]

As Virginia senator Jim Webb put it while introducing legislation to study the situation, "We have 5 percent of the world's population, but we have 25 percent of the world's known prison population. We have an incarceration rate in the United States, the world's greatest democracy, that is five times as high as the average incarceration rate of the rest of the world. There are only two possibilities here: either we have the most evil people on earth living in the United States; or we are doing something dramatically wrong in terms of how we approach the issue of criminal justice."[21] The prison population has grown in parallel with the decline in social connectedness and the rise in economic inequality. Even worse, some 25,000 Americans are now held in solitary confinement, which, as we've seen, can destroy sanity and increase violence.[22]

Then there's screen time. Children under six use screen media including computers, TV, and video-game consoles for an average of two hours a day. This includes babies—for whom even "educational" videos have been shown to reduce language ability. Forty-three percent of children under two watch television or videos every day and a quarter of them have their own TV in their bedrooms.[23] Two-thirds of children under six live in a household where the TV is on more than half of the day—even if no one is watching. In one-third of households with young children, the TV is always on.[24]

All of this adds up to a lot more time spent with things, distracted by media, and in isolation—and a lot less time face-to-face with people—building relationships. Thankfully, there is some good news among these major trends toward social isolation and increasing situations that act against empathy. Time diary studies show that parents today spend more time with their children on average than they did in the 1960s.[25] And increased time doing homework accounts for some of the fall in free play. Rates of volunteering have increased recently. Some studies are also starting to find that social media can increase relational richness by linking friends and family over distances and creating new connections, when used as an addition to rather than a replacement for in-person contact.

There are some qualifiers here, too, however. Parents' subjective experience reported in polls is that they feel like they don't get to spend enough time with their children—and they say that they spend much of the time they are together multitasking.[26] While more time spent on homework sounds good, much of what is assigned isn't actually linked with increased academic performance. Volunteering is a requirement in many schools or seen as essential for college admissions purposes—so it may not necessarily represent a true increase in engagement. And social media can create relational conflict and enable isolation from real connections, not just enhance friendships.

If we all shrink our real social contacts down to just family—as the confidant data suggests we are doing increasingly—we will maintain bonding social capital but lose the bridging kind that connects us to our fellow citizens and allows the government and the economy to function. If we isolate ourselves and don't trust one another, we will miss many of the experiences that not only make us more empathetic— but also make us happier and healthier. If we spend most of our time in front of screens and in human contacts that are commercial or primarily work focused, we get less practice at genuine connection. Societies where people only feel safe with family members and friends aren't very pleasant places to live. As we saw in Chapter 12, they tend to have high levels of corruption, crime, violence, and conflict.

So America does have a real and growing problem with relational poverty, a decline in the circumstances and situations that enhance empathy. And we believe that this is responsible, at least in part, for many of the current health, political, social, and educational challenges

that we face. But we also think that understanding the biology underlying human interactions that we have explored in this book can suggest child-rearing, educational, and public-policy alternatives that
could address these issues. We will explore some of them here.

As we've seen, the choices we make in our childrearing practices,
values, and beliefs can influence how empathy is expressed. Cultural
evolution is ongoing, raising a crucial question for parents: what kind
of children do you want to raise? More broadly, we must also collectively ask: what kind of society do we want to live in? We know that
when we change the way we raise and educate children, when we
change the way we govern ourselves, tremendous cultural transforma-
tions can occur. The United States itself over the last 250 years is a
dynamic example of this. Universal public education and representative democracy—both cultural "inventions"—have led to a remarkably
productive, creative, and socially just society.

Of course, we all see massive room for improvement—and the
signs that the social fabric is disintegrating discussed above are cause
for great concern. Nonetheless, it's important to note that as Martin
Luther King Jr. put it, "the arc of the moral universe is long, but it
bends towards justice." Over the course of history, while the expression of humane and empathetic qualities has waxed and waned, we
believe that some clear progress has been made and it is important to
recognize this if we want to help it continue.

For example, consider the passage below:

> On 2 March 1757 Damiens the regicide was condemned to
> make the "amende honorable" before the main door of the
> Church of Paris where he was to be "taken and conveyed in a
> cart wearing nothing but a shirt holding a torch of burning
> wax weighing two pounds" then, "in said cart to the Place du
> Greve where, on a scaffold that will be erected there, the

flesh will be torn from his breasts, arms, thighs and calves
with red hot pincers, his right hand holding the knife with
which he committed said parricide, burnt with sulfur and on
those places where the flesh will be torn away, poured molten
lead, boiling oil, burning resin, wax and sulfur melted
together and then his body drawn and quartered by four
horses. . . ."[27]

This is the first paragraph of Michel Foucault's *Discipline and
Punish: The Birth of the Prison*—and it includes only a tiny and relatively tame portion of a vivid and nauseating description of a torturous
death. Gruesome torture was not an uncommon form of public entertainment in many societies only a few hundred years ago. Cruel, pro-
tracted killings of both animals and humans were the equivalent of
modern concerts and sporting events. These atrocities were celebra-
ted, the pain of the victims met with laughter and calls for more from
the audience. And they were widely seen as morally acceptable—even
by religious authorities. The torture above, for example, began in front
of a church. The Inquisition—a centuries-long church-sponsored "program" to root out heretics and witches—killed more than 200,000
people and tortured at least twice as many.

As Harvard psychologist Steven Pinker has argued, however, global
attitudes toward torture and many forms of violence have changed.
While the twentieth century was the bloodiest ever in terms of the
sheer number of deaths, over the course of history in much of the
world, violence and torture have actually declined. Pinker writes, "If
the wars of the twentieth century had killed the same proportion of
the population that die in the wars of a typical tribal society, there
would have been two billion deaths, not 100 million."[28] This is not to
downplay the horror of the Holocaust or other genocides—but to simply suggest that there does seem to be a path toward hope.

This idea is supported as well by the long, somewhat bumpy, but still highly significant fall in murder rates. For example, records of homicide have been kept in Western Europe since around the year 1200. All countries show a dramatic decline over the centuries: on average from 28 per 100,000 in the thirteenth and fourteenth centuries to 1.3 per 100,000 in 1975–1994.[29] The current rates are nearly twenty-two times lower than in 1200. (Sadly, the United States lags on this indicator: the 2008 rate for America was 5.4).[30] Though certain periods and particular countries have seen short-term rises, in the long run, the trend is sharply downward, including in the United States. This is good news. Over the centuries, unevenly and with much backsliding and some exceptional regions and periods, our societies have gradually become more humane.

We believe that much of this progress can be understood—and helped to continue—if we recognize the complex interactions between our brains, the cultures that surround them, and the experiences that we provide to our children. As we've seen throughout these pages, there is a bias in our neurobiology that pulls us to be empathic. No matter what kind of structures we invent to live together or to govern ourselves, we continue to carry the genetic predisposition for—and neural systems dedicated to—forming and maintaining reciprocal relationships. Yes, we can and have invented ourselves away from our hunter-gatherer origins—but no matter how hierarchical or oppressive our societies have become, there has always been a human yearning toward freedom, equality, justice, and humanity. This is the powerful biological echo of our ancient and original ways of living together in a relationally rich, minimally hierarchical clan. Though all human societies have inherent conflicts and there are many, many aspects of hunter-gatherer groups that are far from ideal, it does appear that within these early clans, all people inside that magic circle of "us" had value, were involved in key aspects of decision-making, and had core freedoms. This predisposition is likely to be "in our DNA."

And indeed, many of the stated ideals in America's founding documents reflect, enhance, and respect these core neurobiological biases. How well we as a society have succeeded in creating equality, freedom, and justice for all is open for debate—but these values are expressed clearly in both the Declaration of Independence and the Constitution.

In the stories explored here, we have seen how the potential to become humane and empathic is fragile and how, even as we can fully express this capacity with family, friends, and community, we can simultaneously devalue and be inhumane to "others" and "outsiders," however we define them. As we grew from small hunter-gatherer bands to larger clans and tribes, this flip side of our empathic gift— the us-and-them predisposition—started to emerge in powerful and violent ways. Tribalism, unfortunately, is as much a part of our genetic legacy as empathy—and it seems to have been present even when we all had the rich, relational opportunities of growing up in a multigenerational, tight-knit group. The level of violence is indeed very high in tribal conflict.

So what are the best ways to use this knowledge, to allow our complex societies to take advantage of what we know about the brain's developmental needs and the way experience either elicits or eliminates empathy? How can we continue our slow march toward a more complete expression of our species' humanity? What promotes the development of empathy in the individual? What practices, programs, and policies of a society influence the expression of empathy? What do we need to be aware of as we develop ways to cope with our high-tech world?

The first consideration is a very basic one. How much do we really value empathy? Do we really want our children to be empathic, relationally attuned, connected, and interdependent? Do we want them to have well-developed socio-emotional skills? Right now, our policy and cultural priorities don't suggest that we're giving it much thought. In fact, many of the stories here illustrate key influences that interfere with either the development of empathy itself or with breaking through barriers that sustain the current situation in ways that reduce empathy. While most of our children will never—thankfully—experience any of the extremes suffered by some of those whose childhoods we've detailed here, all of them will likely have some exposure to smaller doses of many of these potentially negative influences. The systems that we now have in place for dealing with vulnerable children—like child welfare, foster care, and juvenile justice agencies—often simply don't respect their fundamental neurodevelopmental needs at all.

Though the vast majority of parents would never permit a situation like Brandon's, where a child winds up watching hours upon hours of TV every day, most of our children will spend a great deal of their time in front of some sort of screen. Though few parents would even think of allowing their children to suffer the kind of relational loss that baby Ryan experienced by having eighteen sequential nannies, many of us must contend with childcare situations that are not quite optimal. And although most people would like to have a two-parent family with many emotionally healthy extended family members and friends nearby to support us in child rearing, lots of us do not. A full fifth of American children grow up in poverty, the second highest rate in the developed world.[31] Families unfortunate enough to encounter the mental health, juvenile justice, foster care, and child welfare systems repeatedly discover that these systems create and replicate the chaos, threat, humiliation, trauma, and attachment disruptions that brought their children into these systems in the first place. And the stress of being on the bottom of the social scale in a highly unequal society is not conducive to mental or physical health.

Our children live in a culture of endless cries of "think of the

children"—and attend schools that are crumbling, overcrowded, un-
derstaffed, and short on necessary supplies. They hear again and again
about "family values"—but see their parents struggle to pay for health
care or childcare, often isolated from family and friends and disconnected from neighbors. They hear talk about caring and sharing—but
see around them mainly fear, arguments based on personal attacks,
and competition.

What's worse, empathy and kindness themselves have become
suspect, pathologized even. In the debate that ensued in 2009 after
President Obama said he'd like to nominate an empathetic judge to the
Supreme Court, the chair of the Republican Party ridiculed the idea.
He said, "I'll give you empathy. I'll empathize right on your behind."
No matter what your politics, empathy is clearly one of the capacities
that is necessary in a judge: the golden rule from which much of our
law ultimately stems requires it. And, regardless of political beliefs,
most people will probably agree that such a low level of debate does
not spur civility or help the public understand the issue from both
sides.

Even in the helping professions and psychology—where empathy
would seem to be most valued—caring has come under suspicion. For
example, during the late 1980s to mid-1990s, the "codependency
movement" began to flourish in America, with talk shows, bestselling
books, and related self-help groups attracting the attention of millions.
Starting from the reasonable premise that caring relationships with
addicts and alcoholics can become dysfunctional, this popular strain
of psychology wound up claiming that a large majority of the population is "codependent," and that connecting one's happiness to the happiness of others in any way is actually sick. Being concerned about
others was "taking the focus off of yourself"; caring was seen as an
unhealthy escape. It was selfish, not altruistic. The idea was that only
you can make yourself happy. And, by implication, if your actions hurt

someone else, it's their problem, not yours. If this is true, of course, empathy is superfluous.

While, obviously, people can connect in destructive as well as healthy ways, unfortunately, compassion itself became tainted, almost as though it were an illegal drug. The fact that "dependence" is now the medical euphemism for addiction encapsulates this problem. Addiction experts recognize that even with drugs, it's actually not depending on a drug to function that's problematic (we're all dependent on food, water, and air). Rather, compulsively and destructively using a substance or activity as a way to escape is what makes addiction dysfunctional.

Nonetheless, the self-help, psychological, and psychiatric communities became caught up in the American passion for independence, promoting the idea that true health is needing only oneself, that happiness doesn't require relationships. Some feminists even bought in: "A woman without a man is like a fish without a bicycle." From this perspective, "you must learn to love yourself" first and foremost. Finding yourself—not reaching out to others—is what counts. This combined with the 1970s notion that being brutally honest to others is "helping" them into a noxious ideology that failed to recognize our fundamental need for connection. Consequently, many people believe that expelling a drug-using teen from the family helps the child (while it may be necessary to protect other family members, it hasn't been shown to help children), and counselors and therapists frequently support ending familial or romantic relationships as a solution to conflicts, promoting a "tough love" approach that overwhelmingly prioritizes individual needs. Of course, sometimes it is necessary to end harmful relationships— but this should be a last resort, not a first suggestion.

While the absurd results of the extremes of these views are now recognized, many experts and self-help group members still fail to see that healthy relationships with friends, partners, and family are more important to healing in the long run than pretty much any psychological treatment or medication (though these may certainly be needed to begin and sustain many forms of recovery). It's instructive here to consider schizophrenia, long seen as the most "medical" and incontrovertibly biological psychiatric disorder. Research shows that people with schizophrenia in developing countries are more likely to be in remission from the disease, to show fewer symptoms, to be married and employed, and to be more socially functional than their counterparts in the West—despite having little or no access to medication.[32]

That's right: families in developing countries often do better at helping people with schizophrenia simply by including and accepting them than we do with modern technology. This is not to say that medications can't be useful—it's simply to note that relational health is essential to psychological health. Unfortunately, many Western treatments for mental illnesses and addictions don't reduce—and they can even exacerbate—isolation and stigma. For example, some parents are encouraged by professionals to send nonviolent and nonsuicidal children suffering from grief, loss, and depression to distant residential treatment centers. It's obviously impossible to restore the relationships at home that are essential to long-term healing by exiling the child among strangers—but nonetheless, such treatment is common.

Kindness and empathy have also become devalued due to the intense emphasis on material possessions, status, and wealth common to twenty-first-century capitalism. As psychoanalyst Adam Phillips and historian Barbara Taylor write in their essential book, *On Kindness:*

> Today it is only between parents and children that kindness
> is expected, sanctioned, and indeed obligatory. But before
> we condemn the mother who rages at her toddler in the
> street, we might stop to consider what it feels like to be a

parent in a society where kindness is incidentally praised
while being implicitly discouraged. Kindness—that is the
ability to bear the vulnerability of others and therefore of
oneself—has become a sign of weakness (except of course
amongst saintly people, in whom it is a sign of their
exceptionality). . . . All compassion is self-pity, D. H.
Lawrence remarked, and this usefully formulates the
widespread modern suspicion of kindness: that it is either a
higher form of selfishness (the kind that is morally
triumphant and secretly exploitative or the lowest form of
weakness (kindness is the way the weak control the strong,
the kind are only kind because they haven't got the guts to be
anything else). . . .

  Most people, as they grow up now, secretly believe that
kindness is a virtue of losers. . . . [I]t seems peculiarly
difficult for us to hold on to the fact that we get powerful
pleasure from our own acts of kindness.[33]

Empathy—and the kindness that it enables—is seen as though it's
a health food that doesn't really taste good, something we pretend to
like in hopes of encouraging better behavior in ourselves and our children. We don't act as though unselfishness could be fun or pleasant.
We might look at Mary Gordon's stories about her parents' inclusion of
their children in volunteer work and see it as harsh or depressing, instead of uplifting as she actually experienced it. Somehow, we won't
allow ourselves to believe that helping others could be more than a
grim duty. What is actually one of the greatest sources of human happiness has become a kind of consolation prize, something that even
children can tell isn't really seen as being as good as getting shiny new
things for ourselves.

As psychologist Jonathan Haidt sums up in his book on finding happiness, *The Happiness Hypothesis:*

> If you want to predict how happy someone is, or how long shewill live (and if you are not allowed to ask about her genes or personality), you should find out about her social relationships. Having strong social relationships strengthens the immune system, extends life (more than does quitting smoking), speeds recovery from surgery, and reduces the risks of depression and anxiety disorders. . . . And it's not just that "we all need somebody to lean on"; recent work on giving support shows that caring for others is often more beneficial than is receiving help.

We would argue that the genes and personality qualifier in that statement is unnecessary: without relational health, good genes and an upbeat personality can't produce happiness. Almost all of the recent research on happiness points to one major conclusion: the greatest source of joy is relationships. Our connections to others—and especially the loss of them—are also what make us most miserable, of course. However, the most dependable route to happiness is through altruistic and relational behavior.

Still, no one seems to truly accept that the joy of giving goes to the gift-giver. This is seen as a white lie, something we say to make ourselves feel better. We eat the "veggie" of kindness grudgingly, mainly believing that it's really impossible to enjoy it as much as the sweets of possession and power. This could be, of course, because the highly stratified and market-driven world we now live in makes us all a little less able to take pleasure in connection. Like the pups of the low-licking rats, maybe our dopamine hasn't been properly wired to our oxytocin. But such diminution is obviously a warning sign, a harbinger of possibly increased relational emptiness and further decreased empathy to come.

GIVEN ALL OF this, what can we do realistically to build empathy? First, of all, we need to recognize how important it is and how it matters to everyone, not just to parents or people in caring professions. We hope that the information we've presented here has demonstrated why empathy is essential to everything from health to the economy. But like they say in addiction recovery programs, if we don't first recognize that we have a problem, we're unlikely to change our behavior. Right now, we don't value empathy enough.

Empathy is not just a warm, fuzzy feeling—it makes the modern world possible and allows the economy to grow. It's not just a liberal thing: while conservatives may have attacked President Obama for emphasizing empathy in judges, the right wing's opposition to violent and sexual media and its charitable missions recognize and seek to protect this quality. And empathy is not just maternal and female: studies that find gender differences in empathy favoring women tend to be those that cue subjects to the fact that they are studying empathy, apparently making females focused on appearing more so.[34] Empathy allows us to relieve one another's stress and to make one another happy. It is central to human joy. And we need more of it than ever before to deal with the complex challenges created by our rapidly changing institutions, diverse communities, and distressed planet.

SO HOW DO we make the changes that are needed? Some things that reduce empathy—like too much time spent with screens and too little on face-to-face interactions—are relatively easily remedied by individuals and families by simply becoming conscious of the issue and taking small steps to change. Habits are harder to break than they seem—but at least you have control over these aspects of your own life. Other factors—like schools that tolerate bullying and don't provide enough free time and neighborhoods that overprotect children—can be met with community action. But solving the biggest problems— like lack of affordable, high-quality childcare and health care, the absence of paid family leave, and lack of respect for family time in general—requires coordinated social action and a reframing of the politics that has previously impeded their availability. Even harder to change, of course, are more fundamental problems like severe economic inequality and the stress of poverty and low status that results.

Empathy grows in virtuous circles and declines in vicious ones: if you expand the circle of empathy, it's easier to get people to recognize that "we're all in this together" and to support policies that create a safety net for everyone. Alternatively, coming from an atomized perspective in which we are all suspicious of one another reduces support for such policies. Why should we help "them" when they lie or cheat or don't help themselves? What have "they" ever done for me? This kind of thinking leads to even less trust and more division, making collaboration and empathy harder. In the positive direction, of course, trust leads to more trust and more concern for others, which makes policies that reduce inequality and improve the safety net more desirable, which further increases trust. So how do we move from vice to virtue?

NOT SURPRISINGLY, THERE are many ways to promote the development of empathy—we suggested quite a few either directly or indirectly earlier. Individual practices in the family, school and community programs, and government policies can all influence our ability to give and receive empathy. (For those who want specifics on some promising initiatives, we've included several model programs and key policy suggestions in the epilogue.) But here, we'd like to discuss what parents and communities can do themselves.

As we've seen, empathy originates in the family. The major determinants of an individual's capacity to care are the nature, timing, and quality of his or her foundational relational interactions in infancy and early childhood. All of the neurobiological systems involved in empathy, stress regulation, and reward are being actively organized in babies, even before they leave the womb. So as simple as this seems, the first, crucial step in creating a caring child, a future good citizen, is to care for his mother. Pregnant women need to be safe, nourished, and nurtured. They need to be surrounded by loving people who support them and ease their stress.

After birth, mothers need ongoing and consistent relational support in order for them to best create a safe, nurturing environment for their babies. People mention the cliché "it takes a village"—but practicing this is much harder than preaching it. The bottom line is that an isolated mother is a distressed mother. She will be less capable of caring for her infant. If a mother is available, attentive, attuned, and responsive, her child will almost always develop a healthy capacity for self-regulation. We need to do everything we can to support mothers of young children, from making sure their basic emotional and physical needs are met to just hanging out with them and being available when they need us most.

New parents also need education. As we've seen, from the top to the bottom of the economic spectrum in America, there are huge areas of "child illiteracy" and ignorance about normal development. For example, many people unfortunately still believe that responding quickly to crying babies will "spoil" them. Numerous books and popular sources of parenting advice reinforce this idea. In fact, however, the opposite is true. Parents who respond to an infant's signals and provide nurturing, soothing care cannot spoil their newborns. It is always good to comfort a crying baby—infants cannot be loved "too much" so long as you are responding to their cues. Young children do need small, manageable doses of stress appropriate to their age—but these occur naturally in the course of life.

A good way of thinking about the appropriate dose of stress for a child (or adult for that matter) is the following. Each of us has a "comfort zone" in which we feel totally safe. In this zone, we aren't learning much, because anything even a little bit new will cause a tiny dose of stress. If you move from "calm" to "alert," you are prepared to take in new information: it might be a tiny bit stressful or provoke a little discomfort, but you know you can handle it. "Good" stress keeps us "alert." However, stress that is "too much"—too intense, too long-lasting, or some combination of the two—moves us higher on the continuum into "alarm" or even "fear" or "terror." (See Table 1 in the appendix.) In these states, learning will be compromised. Appropriate stress will keep children alert, but not panicked or terrorized. Pushing a child beyond the "alert" state, trying to force a "breakthrough," is much more likely to produce a "breakdown"—and traumatize, rather than teach the child a lesson. We all need emotional safety to maximize our learning potential: everyone learns best when they feel up to a challenge, not overwhelmed.

For the littlest children, time itself is an essential element in creating empathy. We'd all love to believe that what matters is "quality time," not sheer quantity. Unfortunately, as we saw in Ryan's case, that's simply not true. In infancy and early childhood, quality time is only part of the equation—babies need many, many hours every day of one-on-one attention from the same few people over and over in order to build the full relational capacity of their brains. Babies can't generalize until they have learned individual characteristics through repeated experience: they live in the specific repetitions of particular relationships. Simply stated, we need to spend more time around and with our children, particularly in their first year of life. These little ones need to watch us interact with others, they need to hear us, feel our touch, see our smiles. We all benefit when parents can spend this precious and fleeting time with infants.

And, of course, one of the best ways for a parent to spend time productively with a young child is to read to them. The warm, nurturing interaction that take place during this playful, engaged social *and* cognitively stimulating experience leads to a remarkable association. The child comes to associate—to connect—reading with pleasure. The child who learns to read while sitting on Mommy's or Daddy's lap will become a lifelong learner. In contrast, the child who is made to sit still and listen and read in a classroom where there is no tolerance for play, touch, *or* movement during this highly cognitive activity will learn to read—but will often come to hate reading. Given the chance later in life, such children will rarely read for pleasure.

And reading to children is in itself a great way to encourage empathy. Indeed, some historians and sociologists believe that the spread of literacy in many parts of the world actually *caused* the slow decline in murder, torture, and violence seen in the recent history of these places. Reading "builds" networks in the cortex: the area of the brain responsible for planning and impulse control. Greater self-control tends to reduce violence. And reading fiction—particularly novels written in the first person or using exchanges of letters—explicitly requires perspective taking, placing the reader in the position of characters and eliciting pleasure from their triumphs and pain in their suffering. Reading such books is essentially practicing empathy.

Talking with a child in order to elicit perspective taking whenever possible helps, too. When you read to them or discuss books, ask what they think the characters are thinking and feeling. Point out facial expressions and body language and talk about what these mean. Do this with movies, music, video games, television, and real-life examples as well. It's especially important to have explicit discussions around perspective taking, body language, and tone of voice if your child seems to be having trouble with social signals. And doing this via discussions of books and other media is a natural and unobtrusive way to teach and reinforce it.

Promoting perspective taking is particularly important when disciplining a child. The most effective way to get children to behave well is to help them enjoy doing so and to utilize their natural desire to please their parents to encourage this. Needless to say, spanking or any other form of harsh discipline does not and cannot encourage empathy: empathy is learned by having the experience of being treated kindly, not by being made to suffer. People often misunderstand this connection because they see that many people who have suffered empathize deeply with others in pain—like Trinity. They think: if only the bullies could have the experience of suffering what their victims do, then they'd understand. As we've seen, though, most bullies do have the experience of being victimized—and it makes them want to get even, not help others. It wasn't Trinity's suffering that made her empathetic: it was her sensitive nature and ability to connect despite experiencing little empathy at home that enabled her response.

To encourage empathy, discipline by reasoning, perspective taking, consistency of appropriate consequences, and above all, love. Research shows that children who receive corporal punishment are more aggressive, more likely to be antisocial as teenagers, and may even have lower IQs than those who are not physically disciplined. Ninety percent of the research on spanking shows negative effects. While some studies have found improvements in discipline linked to spanking in African American culture where it has traditionally been more accepted, the study that looked at IQ found negative effects across race.[35]

Of course, some of the connection between corporal punishment and misbehavior could be due to the fact that more difficult children are likelier to be spanked—but even if this is true, it means that the spanking doesn't solve and may even exacerbate the problem. One of the most recent studies followed over seven hundred children from age five to fifteen or sixteen. Unsurprisingly, those who were spanked showed more problems in their teens and had worse relationships with their parents—but the connection remained even when the researchers adjusted statistically for early misbehavior.[36] Using your own words helps children to use theirs when they are tempted to be aggressive. More responsive, verbal, and sensitive parenting improves not only behavior, but academic performance. If you teach children to behave by using reason, they are likelier to be reasonable. It is true that using reasoning can make children more likely to challenge your rationale for telling them what to do—but remember, children who challenge you will also be more likely to challenge peers when they do something wrong, rather than just doing as they are told by others.

As children get older, more explicit discussions with them can also promote empathy. This can include talking specifically about how people's actions affect other people and the reasons why other people might have different reactions and points of view. Encourage children to think about the less fortunate and how they can help. Expose them to a variety of perspectives. Eat family meals together. Consider trying the strategy that Mary Gordon's parents did: limit dinnertime discussion to ideas, art, and literature. Making the table a place where you talk about values and what really matters can bring everyone closer—and also provide a refuge from discussions about discipline or other personalized points of contention. Remember that pleasure usually teaches much more effectively than pain: this is why addicts have difficulty *stopping* taking drugs, not starting to do so.

After infancy, of course, a child still needs access to high-quality caregiving: this is a major challenge in our culture today. Unfortu-

nately, many of our childcare environments are relationally impoverished, with inadequate ratios of caregivers to children. Parents should
make sure that childcare settings do not use televisions as babysitters;
they should ensure that primary staff is assigned to mind individual
children and that staff members have adequate training. Parents should
also choose kindergartens and other childcare settings that do not over-
emphasize cognitive development at the cost of socio-emotional learning. Three-, four-, and five-year-old children need to have lots of time
for unstructured play. They need to explore, to negotiate the rules of
made-up games with friends; they need opportunities to practice com-
promise, negotiation, and sharing.

Three-year-olds should learn to read in the laps of caring parents,
siblings, or grandparents—not while being forced to sit in a "big,
apple-pie circle." They will learn more easily and more naturally in the
richly relational setting of the home. Clearly, childcare issues alone
could encompass an entire book—but we want to stress here that
early childhood is a crucial time of life for the development of empathy. A greater society-wide focus on providing adequate nurturing,
modeling of healthy relational skills, and simple access to relationships during early childhood is essential if we are to create a caring
America. Programs, policies, and practices that target older children
and teens will all be playing catch-up if we don't take advantage of the
window of opportunity provided by the intense malleability of our
young children's brains.

It's also important to recognize that play is at the heart of developing empathy. Unfortunately, many children now spend so much time
in highly structured, adult-dominated activities and in front of video
screens that spontaneous, self-motivated play is becoming rare. Minimizing screen time and maximizing the time available for free, unstructured, and outdoor play is essential for healthy socio-emotional
development. Concerned parents who want to organize others can

help not only increase connectedness for both children and adults—
but also enable more outdoor play by supporting one another in letting
children play more freely.

Schools also need to be engaged and helped to become more developmentally aware. It is remarkable how many of the elements of
modern education decrease opportunities for healthy relational interactions and by doing so actually undermine the core mission of education. The way many schools structure their classes and other activities
can actually make it harder to learn both cognitive and socio-emotional
content. One primary structural flaw of the modern school is age segregation by grade. This minimizes opportunities for modeling and interacting with others who are older or younger. Obviously, some types
of learning are easier in age-graded groups—but we need to make
room for contact across ages, too.

A heterogeneous clustering of students (similar to that seen in a
one-room schoolhouse) would lead to a much richer relational milieu
that would promote empathy. Indeed, the fundamental structure of
Mary Gordon's Roots of Empathy is to bring some element of heterogeneous clustering to the classroom by allowing older children to see
and spend time with mothers and babies.

Schools also need to develop policies that preserve recess, lunch
periods, gym classes, and other "down time." This will not take away
from academic performance. Like adults, children need breaks and
time to socialize. Ensuring that the homework burden is not too heavy
is another step to take: research does not link high amounts of homework with better test scores; in fact, too much actually reduces performance in math.[37] Play is indeed a big part of learning empathy:
research shows that poor children who have less structured activities
are actually more self-reliant and caring.[38] It is quite possible to overdo
an emphasis on cognitive achievement. Of course, children whose home
environments include little cognitive stimulation may need longer

school days, after-school programs with tutoring, and opportunities such as open gym, or summer programs—but even here, there are biological limits that must be respected.

Emotional safety at school and in the community is necessary, too. Remember that learning can only take place when children are in the "calm" or "alert" region of the arousal spectrum, when they are not amped up on stress hormones because they're worried about being beaten up in the hallway. A girl anxious about being sexually harassed because her breasts have developed early is not one whose cortex is functioning optimally to allow her to remember the stages of mitosis; a boy who fears he will be shoved into a locker during gym class isn't going to be focused on verb tenses.

Recess and gym and other periods of down time can be hell for children in schools and communities that tolerate bullying. As Mary Gordon points out, one of the biggest reasons children play sick or skip school is to avoid bullies: making schools safe and inclusive can boost academic performance not only by increasing attendance but by decreasing uncontrollable stress and therefore allowing children to focus better on schoolwork. Virtually all of the schools that have been sullied by shootings or by sexual assaults like Ryan's and those in Glen Ridge have had a school culture that turns a blind eye to acts of aggression and social intimidation by "popular" students, especially athletes. The adults who run these schools don't believe that they can challenge the "natural" social hierarchies that children develop. Rather than trying to do so, they instead affirm them and covertly—or sometimes even overtly—side with the "winners" and participate in excluding "losers." This makes bystanders afraid to intervene when bullies victimize people; it creates a culture of impunity. The children who want to be empathetic are forced to hide their concerns and avoid speaking up for victims for fear of becoming the next target.

In contrast, schools that work from the top down to be inclusive and to set a tone that discourages bullying and promotes cooperation can make a real difference. There is a wide range of school climates all across the economic spectrum—as the Glen Ridge case shows, schools that tolerate bullying are not limited to inner cities with over-stretched education budgets. And warm, caring schools are not found only in rich suburbs.

While inclusive schools can't eliminate social hierarchies or eradicate bullying, those that make a full-court press to minimize related problems can achieve remarkable results. Studies find that children feel more connected to inclusive schools. And schools that report greater student connectedness have lower rates of drug use, violence, heavy drinking, smoking, and suicide attempts—so the benefits don't just go to the children who would otherwise be victimized, but to the whole community.[39]

Another important factor is ongoing exposure to different types of people. Diverse, multicultural schools and communities can help children become familiar with people of other races, socioeconomic classes, religions, and cultures. Familiarity is a great way to increase empathy: as we saw in Wendell Potter's story, it's much harder to ignore the suffering of people you grew up with than it is to dismiss the woes of strangers. If you know that people of different races and classes are similar to you in many ways—if you can't characterize them as alien—it's much easier to get support for policies that extend help to everyone, like more school spending and universal preschool or health care.

It's important to manage diverse schools carefully, however, because of the way children form social groups. Children often self-segregate, gravitating to people who seem most like them on the surface and often producing us-versus-them hostilities that reduce the tolerance and acceptance that diverse schools seek to increase. Creating activities, classes, and organizations that appeal to all types of children, and working to minimize self-segregation when possible can help. As we learn more about social-network formation and how groups work, we will be better able to create mixes of students that work for everyone.

WRITER MALCOLM GLADWELL opened his book *Outliers* with an unusual introduction. It tells the story of a village in rural Pennsylvania that was, at first, a medical mystery. In the late 1950s, a local doctor observed that among his patients from local communities, he almost never saw a man under sixty-five from Roseto, Pennsylvania, with heart disease. Since cardiovascular disorders were then, and still are, the number one killer of men under sixty-five—and at the time, there weren't many effective treatments—another doctor decided to investigate. That man, Stewart Wolf, and sociologist John Bruhn began a collaboration that would last decades. They first discovered that no one in the village under fifty-five had died recently of a heart attack. None of the men even showed any sign at all of heart disease: not elevated cholesterol, nothing. Despite having similarly high rates of smoking and high-fat diets, Roseto residents didn't experience cardiovascular disorders at anywhere near the rate of the neighboring towns that the researchers used as control groups. Men over sixty-five had a 50 percent lower rate of heart disease, compared to American men in general. Addictions, depression, suicide, crime: all rates were low or nonexistent. Overall, the mortality rate in the town was 30 to 35 percent lower than the national rate.[40]

If you've been with us this far or have read Gladwell's book, you can probably recall or guess what made Roseto healthy. It wasn't red wine or other dietary elements. It wasn't genetics or exercise. It wasn't a local spring that provided excellent mineral water. It was, indeed, the town's close-knit families and their relationships with one another. In Roseto, many families lived in multigenerational households: children,

parents, grandparents, all together. They ate many family meals together; there were no fewer than twenty-two separate civic organizations in a small town of only about two thousand people. As Wolf and
Bruhn described Roseto:

> Their way of life emphasized cooperation and sharing rather
> than competition. The town radiated a kind of joyous team
> spirit as its inhabitants celebrated religious festivals and
> family landmarks such as birthdays, graduations, and
> engagements. Their social focus was on the family, whereas
> neighboring communities, holding to the traditional
> American view, were more likely to focus on the individual as
> the unit of society.
>
> When we interviewed and examined inhabitants of
> Roseto and familiarized ourselves with the town and its
> people, we were surprised at our inability to distinguish by
> dress, manner, or speech the affluent owners of textile
> factories from the more impecunious laborers. . . . The lack of
> display of affluence or even obvious distinction between rich
> and poor, and the absence of need to "keep up with the
> Joneses" appeared to be a central ingredient in the unifying
> cohesive force of the community. We learned that it had its
> origin in the dim past of feudal or prefeudal Italy, mainly from
> the myth of *maloccio,* the evil eye. The belief was that any
> ostentation or display of superiority over one's neighbor
> would be punished by ill fortune.[41]

So there you have it: as we've seen again and again in this book,
deep relational connections, reduced emphasis on status and inequality, and a culture that promotes cooperation rather than competition:

all of this dramatically increases in profound and measurable ways not just mental health and happiness, but physical health. These "warm and fuzzy" qualities helped Rosetans live years, even decades, longer than the more isolated people in the towns around them—indeed, they lived longer than people born in Roseto who had moved away.

Gladwell closes his introduction by saying, "I want to do for our understanding of success what Stewart Wolf did for our understanding of health." In other words, the goal of his book was to demonstrate that "it takes a village" to make a genius or otherwise successful "outlier," just as it takes a village to keep people healthy. Both are noble missions. Unfortunately, in reality, Wolf's work has had almost no influence on medicine. We still talk mostly about diet, exercise, and medication in the prevention and treatment of heart disease; doctors don't say "form two new friendships and visit your aunt Jane more often" to those with cardiovascular disorders. Even psychiatrists—who clearly recognize the role of relational health in mental health—often do little to promote more and better relationships among the patients they treat. We believe relational health needs much more respect and consideration.

Of course, no community, modern or ancient, gets it completely right: in fact, Roseto's health advantages melted away over the years as people moved away or became more assimilated into the American individualism around them. We don't believe there has ever been a golden age of relational health. While, as we've seen, hunter-gatherer societies do have a relational richness that many modern cultures don't, they also have higher rates of intergroup violence and infant mortality that no one would wish to replicate. Our groups and families can clearly harm as well as heal. Our human contradictions make perfection impossible. Nonetheless, we hope we've helped you think about relational health—and how improving our connections to one another through greater empathy could make a major difference in

both direct and indirect ways. We think we could use more community, cooperation, and connection, particularly in light of the rapid
technological changes we face.

Will increasing empathy solve all the world's problems? Of course
not. But few of them can be solved without it.

epilogue | PEOPLE AND PROGRAMS

epilogue | PEOPLE AND PROGRAMS

M ANY OF THE PEOPLE WHOSE STORIES appear here have thought a great deal about ways to increase empathy in the world. We would like to highlight some of their efforts—as well as a few exemplary programs and initiatives here. The most important aspect of any effective program is that it work to increase the number and quality of relationships in the lives of children and families. Recognizing this and highlighting other critical components can help those who want to work toward this goal find good models and ideas to guide them. And the individuals who do this work deserve recognition as well.

EUGENIA IS PLANNING to study psychology or another field that examines the human mind, hoping to understand more about child neglect and early development, and prevent other children from experiencing the treatment she and her brother received in orphanages. She will work to understand and magnify empathy through her education and career goals. And she continues her volunteer work with animals.

SAM HAS FOCUSED on advocacy and activism around autism spectrum conditions, trying to destigmatize them and increase public understanding. He wants to help the nonautistic world recognize that autism is not just a "disability" or "handicap," but like many other "different" ways of thinking carries important gifts as well—gifts that we ignore at great cost to everyone.

As he said in an online interview, "Autism does not *have* to be a tragedy. Stop. Think. *Listen* to autistic people themselves. Read and view what autistic self-advocates have written and filmed. Understand that there are self-advocates at all points on the spectrum, not just so-called high functioning. Understand the distinction between handicap secondary to autism, and autism per se—then go forth and mitigate the handicaps, nourish and support the autistic ways of thinking, sensing, feeling, and encountering the world, and help your autistic loved one grow into living fully and capably as an autistic person."

Sam stresses that there are difficulties with empathy on both sides of the autistic divide. Part of the reason autistic people have problems connecting is that many people don't understand their sensory issues or the way their minds work—just as autistic people can wrongly project their perspectives and misunderstand everyone else. One of his goals is to help both sides appreciate and connect with each other better.

TRINITY WALLACE-ELLIS'S APPROACH has been to lead by example and effort, showing the foster children she works with that another life is possible and that giving and nurturing is rewarding. Her warmth and her refusal to abandon her family members despite some ongoing difficulties stand in sharp contrast to those who say that "tough love" is the only way. Trinity was even able to reconcile with her father: eight years before he died he succeeded at quitting heroin, using support from a twelve-step recovery program. That program suggests making amends with those you have harmed during your addiction. Realizing that much of what he'd done was irreversible, he nonetheless accepted responsibility for his actions and tried to behave differently in recovery, and Trinity found some peace in that.

Trinity has spent her career fighting to improve the lives of children in foster care. Through the California Youth Connection, which she joined at sixteen, she testified and named the legislation that the state passed to try to keep siblings together in foster care and improve contact between them if this was not possible. At nineteen, she began working for the Los Angeles County Department of Children's Services in a program aimed at making the system more responsive and simultaneously providing employment for children who'd been through foster care. The idea was to help them become social workers—but as is often the case, budget cuts meant that this didn't happen. Trinity moved on to other agencies and continues to work in various ways toward improving the childhoods of the most vulnerable youths. She currently runs her own child services consulting firm—not surprisingly, it's called the Resiliency Group—and she volunteers as an adult supporter of California Youth Connection and for the Foster Care Alumni of America.

MARY GORDON WORKS one child and classroom at a time, using the relationship in which empathy is born to demonstrate it in Roots of Empathy (ROE), which we visited in Chapter 1. ROE models key principles for learning empathy that can be used by anyone who wants to be more empathetic or help their children become that way. Consequently, we want to describe its approach in more detail here.

ROE is conducted in twenty-seven sessions over the course of a school year. Nine of these include the baby, and they are roughly a month apart. In each of these "family visits," a parent and infant visit the class, and the baby is placed in the center of the room on a green blanket, interacting with the parent and the other children. There are separate curricula for kindergarten, grades one through three, four through six, and seven through eight, which are tailored to children's interests and abilities in those grades. After the family visit, a trained ROE instructor conducts a follow-up lesson in the regular classroom, accompanied by the classroom teacher.

Gordon says, "When we talk about the deepest communication, it's rarely verbal. It's in the looks and the sighs and the touch. Babies have all of those; their bodies are theaters of emotion." In fact, ROE children actually often learn more from the program if the baby is colicky and cries a lot. "It's wonderful when we get a colicky baby," Gordon says—and she's not being sarcastic. She adds, "We all agree that [the crying is] totally annoying. We ask the mother how she feels or the father, whoever is there. They have an opportunity to quite honestly say that they cry themselves sometimes because they're so frustrated."

Children rarely consider the world from the parents' perspective—and parents themselves often inadvertently discourage this by trying to hide their insecurities and fears from their children to protect them. "Children never get to hear how an adult feels around a negative emotion," Gordon says. "Roots of Empathy parents share all that with children." By doing so, they let youths know that their own fears and emotions are normal and that it's OK to express them. Of course, it's probably easier for children to empathize with a parent who isn't their own mom or dad. That way, the emotional complexities of that relationship don't intervene.

Dealing with a colicky baby also allows the instructor to teach about temperament, one of the most important lessons imparted by ROE. "A crying baby is not a bad baby," says Gordon. "This is a baby with a problem. We don't necessarily know how to help the baby and we're going to try a variety of things. We're going to watch and see what helps." This lesson alone may help reduce child abuse by teaching children about how to deal with babies in a neutral setting before they ever encounter such a situation as parents themselves. It's possible that ROE may also work to prevent teen pregnancy by showing just how much is required to be a good parent—but this hasn't yet been studied.

The simple exposure of children to nurturing caregiving can make a deep impression. Gordon describes an interaction between one ROE baby and a thirteen-year-old boy whose mother had been murdered when he was four. He'd been in multiple foster homes and was often aggressive and disengaged with school. He looked intimidating, with a shaved, tattooed head. At a family visit, the mother offered the children a chance to hold the baby—and this boy volunteered. The mom must have been anxious, but nevertheless she handed the child to him. He tenderly held the baby, who snuggled right up to him, burying her face in his shoulder. He rocked her quietly and reverently. Afterward, he asked the instructor, "Do you think if nobody ever loved you that you could be a good parent?"

ROE imparts powerful lessons about parenting, both implicitly and explicitly. The children are encouraged to ask questions and even to ask permission from the baby before they do anything involving her. They observe the child's response before and after taking any action, learning that everyone's feelings need to be respected, even if they can't express them verbally. They see how babies turn away or wriggle when something they don't like is happening, even before they cry. ROE children are also taught specifically never to shake a baby, a lesson that they seem to take home and share with their parents.

Understanding that babies have different temperaments and that none of these is all "bad" or "good" also helps the children understand their own temperaments and personalities. Gordon describes a baby who is extremely unhappy when a toy is taken away. "This little baby is really persistent and intense, very challenging to raise, and she makes her mommy very tired sometimes," she says, describing what she would tell a class. "But when she gets to be your age in school, she will keep persisting and she will be intense as an adult. And that's a good thing," she explains, noting that such children not only cry but also laugh a lot and can motivate themselves and others. Children are rarely taught explicitly about temperament—but knowing that people can have naturally different styles and aren't bad or wrong for being more or less active or more or less persistent often helps them adjust and be more tolerant of both themselves and others.

ROE also teaches tolerance of other differences. The program seeks specifically to include parents from varied ethnic backgrounds and nontraditional families. It also seeks babies who are themselves "different"—infants with Down syndrome or a cleft palate, for example. One baby Mary remembers particularly had a large birthmark—not the same kind as Jeremy's, but one that was large and visible on the top of her face. The mother was anxious about participating because she'd already seen people have negative responses to her baby. Gordon briefed the teacher and children—who were in second grade—in advance. "A lot of them wanted to know about the baby's feelings. 'Did it hurt her?' Right away there was that connection; they began to fall in love with that baby."

Soon, however, the children were comparing their own birthmarks, and the teacher had to ensure that they didn't strip to show off birthmarks under their clothes. In fact, the boys and girls now thought it was cool to have a birthmark, and the ones who didn't have any felt a little left out. The mother was happily surprised. This was not at all what she'd expected. When a new boy joined the class later in the year, the other children told him about the birthmark matter-of-factly and their acceptance kindled his.

Of course—as is clear from some of Gordon's stories—there are serious limits to what short-term, school-based programs alone can do. Consequently, there are several larger-scale initiatives aimed at increasing empathy and relational connection that we'd like to mention. Not surprisingly, many of these focus on improving the early childhoods of the poorest and most vulnerable children.

IN NEW YORK, Philadelphia, and Baltimore, efforts are being made to work across multiple systems—schools, preschools, high schools, child welfare, criminal justice—to help the most distressed families and children. Labeled "Promise Neighborhoods" by President Obama, these programs provide multiple child-focused services intensively in one particular area. All begin with the premise that prenatal care and support during the first years of life provide a critical foundation for later life success. Keeping in mind that by preschool, children of middle-class professionals know twice as many words as children raised in poverty, these initiatives seek to change this by enriching the early environment and creating schools and preschools that support these efforts. Some seek to coordinate the multiple systems that may interact with a family or child to avoid, for example, having a mother who is seeking treatment for an addiction lose custody of a child due to lack of childcare when she has therapy sessions.

At the very least, these programs seek to reduce childhood problems—like Ryan's, although he was in a very different socioeconomic class—that are caused by parental ignorance. Recognizing that poor parents have strengths as well as weaknesses, these programs also work to build on them. For example, research shows that the parenting styles used by poor parents encourage more free play and independence: elements of childhood that have been dramatically reduced in the middle and upper classes, to the detriment of children's social development. The idea is to take what works best from all sources—and empower parents to utilize it.

The best-known program is Harlem Children's Zone. Founded by Geoffrey Canada, who himself grew up in poverty in the South Bronx and ultimately received a graduate degree in education from Harvard, it is described extensively in *New York Times* reporter Paul Tough's book, *Whatever It Takes*. Harlem Children's Zone now encompasses ninety-seven blocks in northern Manhattan. It serves about 8,200 of the 11,300 children living in the zone in at least one of its more than twenty programs. Nearly three-quarters of children in the neighborhood are born in poverty.[1] Starting with "Baby College," which teaches new and expecting parents important facts about child development, the Zone includes high-quality daycare, preschool, outreach workers who do home visits, and after-school programs for teens. It also runs schools reaching from kindergarten to high school. About 1,500 employees staff the various programs—also increasing local employment, although the recession has caused a 10 percent staff reduction. Most of the funding so far has come from private sources.

Canada wants to engage the whole neighborhood: helping families from pregnancy to college acceptance. Although the program includes charter schools, unlike most such schools, Harlem Children's Zone schools do not admit only the best students or those with the most committed parents. They are open to everyone in the neighborhood, via a lottery system that gives precedence to those who have participated in the early childhood aspects. These children are selected via vigorous outreach to ensure that the program isn't just cherry-picking children who would be most likely to do well anyway. Canada's goal is to give poor children the advantages and cultural expectations of academic achievement that more often mark middle-class life.

Regarding discipline, Canada told Tough, "A lot of the families we work with believe good kids are quiet kids. If you're a good parent, your child listens to you, if you're a bad parent, your child doesn't. Well, the problem is that *no* two-year-old listens to a parent. But no one has ever explained that to a lot of our parents. So you see parents smacking a two-year-old's hand saying, 'Didn't I tell you not to do that?'"[2]

Like Roots of Empathy, Harlem Children's Zone programs use research about brain development to help these parents and children benefit from it the way middle-class parents have. Writes Tough: "The middle-class style of discipline—negotiation, explanation, impulse control—was intertwined with the middle-class style of brain development. There was simply more talk in middle-class discipline, and thus more verbal stimulation; encouraging a child to understand the reasons for a prohibition, and to take part in choosing an alternative, was a powerful cognitive stimulus. If you follow that path, the instructors told parents, your child will be smarter and happier and will make better decisions later on."[3]

As Harlem Children's Zone has grown, test scores in its schools have improved—not always evenly, but measurably. Math scores in the middle school have eliminated the racial disparity in performance in the city—they are now equivalent to those for white students. Five hundred and fifty participants in the after-school programs are headed to college in fall 2009—a high number from a neighborhood that typically sends few students to college.[4] As the children who have been most heavily involved with the program from early life advance through the grades, we expect to see much greater changes—as well as measurable reductions in aggressive behavior and crime.

Inspired by the success of the Harlem Children's Zone, President Obama plans to fund Promise Neighborhoods in twenty cities across the country. Ten million dollars has already been budgeted for 2010 to start the process.[5] We believe that these projects could significantly improve the lives of thousands of the most needy children and families.

ONE ELEMENT COMMON to most Promise Neighborhoods involves another highly regarded early childhood intervention: home visits by trained nurses to women during pregnancy and the child's first few years of life. These visits are integrated into the Harlem Children's Zone project and extensive research has shown that they get results. David Olds, a professor of pediatrics and prevention at the University of Colorado, founded the program in the 1970s. New mothers living in poverty—often single, teenage, on welfare, and struggling with additional problems like addictions and domestic violence—are visited by nurses once a week during pregnancy and the first few months of the baby's life. Most women receive about sixty-four visits in total—which become less frequent as the child grows and typically end when he or she is about two and a half.

Each nurse acts as a combination of coach, teacher, surrogate grandmother or older sister, and baby nurse: they provide important information about child development, discipline, and health. They especially emphasize elements that research has found to be important in school readiness: to reduce the vocabulary gap between middle-class and poor children in language development, for example, they teach mothers to talk frequently to their babies, not just yell at them when they do something wrong. Based on the research showing that poor children hear more "discouragements" when spoken to than "encouragements," they make an effort to change that ratio. Essentially, they promote nurturing, responsive care. They also provide crucial support for the mother, teaching her that, for example, if a baby won't stop crying, it doesn't mean that she's a bad parent. As Olds told Maia,

"Learning to understand children's motivations and abilities helps parents treat them more sensitively and responsively, and that makes it easier for children to accept guidance and not respond provocatively."

The results are impressive. One study, published in the *Journal of the American Medical Association* in 1998 found a 59 percent reduction in the arrest rate of teens whose mothers had been visited by nurses compared to those who were not, some fifteen years after the visits had taken place. Another found a 50 percent drop in substantiated reports of child abuse or neglect in families who received nurse visits. Visited mothers tend to space their pregnancies farther apart and to suffer less prenatal hypertension, which can interfere with fetal development. A 2005 study conducted by the Rand Corporation found that for every dollar spent on nurse visits to vulnerable families, six dollars could be saved in future welfare, health, and juvenile justice costs. President Obama has called for the program to be expanded to reach every first-time mother living in poverty and made funding for it part of his health reform bill, which is being debated as of this writing. The Senate has currently appropriated $1.5 billion for state grants to fund the program.

EARLY CHILDHOOD IS not the only part of the lifespan that affects the development of empathy in the United States. We are not only a "child illiterate" society, but an "elder illiterate" one as well. As we've seen, it's hard to empathize with kinds of people with whom you have little interaction: if you don't know how two-year-olds are supposed to behave, it's hard to know how to connect with them. Similarly, we have become disconnected from older people. While many people once had multiple generations under one roof, now many older people retire in distant communities. Some even live in retirement villages where families with children are explicitly excluded from residing. Though most people seek desperately to avoid living in a nursing home, they often fail and wind up surrounded only by those paid to care for them and other elderly people in the same situation.

This age segregation reduces relational opportunities for everyone. It contributes to fear of aging and to the marginalization of the old. It robs us of a resource—and produces yet another self-perpetuating cycle. In this one, old people are seen as irrelevant, rather than as sources of wisdom, merely because of their age. The lessons of their lives are lost because their relevance is discarded without consideration—and this leads to further dismissal. Consequently, opportunities for intergenerational relationships are lost.

There are two programs that we'd like to highlight that are working to fight this. One, San Pasqual Academy near San Diego, offers older people reduced rent on their homes in return for serving as "foster grandparents" to nearby children in foster care who live at the academy. By bringing these groups—both of whom are often isolated and lack rich relational networks—together, both sides benefit. Generations of Hope in Rantoul, Illinois, is similar: there, families with foster children live near older people who serve as surrogate grandparents. The grandparents also receive subsidized rent. When these relationships click, they last forever. One Rantoul resident, herself childless, told the *New York Times* that she felt that this was the most important work she'd done in her life and that she was so happy she felt like "a cowgirl on wheels around these kids."[6] Given the health benefits we now know come with relationships, we believe that more initiatives like these should be started and supported.

ACKNOWLEDGMENTS

From Bruce D. Perry

The work that I do as a child psychiatrist brings me into contact with a remarkable range of people. Some are damaged by the turmoil of their tragic lives—and some of these damaged children grow up and go on to damage others. Yet most of the people I meet have found ways to heal and move forward. You've met some of them here: Eugenia, Trinity, Ray. I am grateful for their generosity and courage. There are so many more who remain unnamed and unrecognized. Each day at the ChildTrauma Academy (www.ChildTrauma.org), we receive e-mails and calls from professionals and individuals sharing their experiences. When we hear their stories, we always learn; almost always they give us encouragement to keep learning and sharing our work. I am thankful for all of these generous people and the small but powerful relational rewards they provide. They keep us going as we do this challenging work.

Special thanks are due to my colleagues at the ChildTrauma Academy: executive director Jana Rosenfelt, director of programs Dr. Christine Dobson, and director of education and training Emily Perry. This small group is amazingly productive, especially when I leave them alone. The ChildTrauma Academy fellows all across the world continue to provide me with new ideas and help challenge my old ideas. They are a remarkable group. Special recognition should go to Kristi Brandt, Rick Gaskill, Gene Griffin, Jerry Yager, Annette Jackson, Kalena Babeshoff, Mary

Beth Arcidiacono, and Diane Vines, all of whom are active collaborators
in our core research and training activities at the CTA.

There are several remarkable institutions working with us at the CTA
that I should acknowledge as well. At these sites, children and families are
served, and the professional staff continues to help us understand how
maltreatment damages and how empathy heals children: Sandhill Child
Development Center in Las Lunas, New Mexico; Youthville, United Meth-
odist of Kansas; Mount St. Vincent Home in Denver; the Denver Children's Home; Alexander Youth Network in Charlotte, North Carolina; the
Oprah Winfrey Leadership Academy for Girls in South Africa; and Take
Two of Berry Street in Melbourne, Australia.

Time to think and time to create is hard to find. Special support for
creative and productive time comes from the generous support of Sheila
Johnson and the Amon Carter Foundation, and the ongoing support of
Dick and Meg Weekley.

This book is the second that I have had the privilege to write with
Maia Szalavitz. There is no doubt that she is the engine behind this book.
While I am off involved in the ongoing work of the ChildTrauma Academy,
she is writing, researching, interviewing, and doing all of the hard work
necessary to write a book. She is patient when I'm unavailable and great
at begging editors for more time when I'm just too busy to write as much
or as fast as we need to, to meet deadlines. Maia has a gift for taking fac-
toids and helping transform them into knowledge. And while the range of
her knowledge is remarkable, it is nothing compared to the range of her
compassion. She has a big, big heart. This makes for a powerful and positive combination. I am lucky to be in such a wonderful collaboration.

I am always most thankful for my family. There is nothing I have ever
done that would be possible without their love, support, and tolerance. So
to Jay, Emily, Maddie, Elizabeth, Katie, Martha, and Robbie, thank you
for helping me love, laugh, and learn. Finally, I would like to thank my
wife, Barbara, who is the bedrock of the family and sustains us all. I am
only able to go off and do all manner of interesting things because she is
so solid, strong, and present at home. She is the one who should be writing books about children and empathy; she knows way more about it
than I do.

From Maia Szalavitz

Writing a book about empathy and the interdependence that it reflects highlights one's own connections to and reliance on others. First, I'd like to thank my coauthor for once again being a delight as a collaborator and for making me a genuinely equal partner in this work: I truly appreciate your generosity. I am honored to continue to work with you. Second, I'd like to reiterate our thanks to everyone who shared their stories with us and helped us tell them here: Mary Gordon, Sam, Trinity, Sue and Eugenia, Ray, Asdis Omarsdottir, Mary and Sophia, Shannon Keating and all of her wonderful sixth graders, and Roots of Empathy curriculum director Donna Letchford.

Alissa Quart, Rachel Lehmann-Haupt, and Deborah Siegel were wonderful and supportive early readers: thank you "Matilda's"!!! Thanks to Annie Paul, too, for a great read and inspired suggestions—and to my mom, Nora Staffanell, who was another important reader. Lisa Rae Coleman, I couldn't have done it without your transcription and friendship: I want to be in *your* book acknowledgments soon (same to my friends Anne Kornhauser and Cynthia Cotts, minus the transcription part)! Additional gratitude to Jessie Klein and to my editor at *Time* magazine online, Sora Song. Thanks to Trevor Butterworth and Don Rieck of stats.org for all your support and encouragement!

We couldn't have done it without our fabulous agent, Andrew Stuart, and gratitude is due as well to Sarah Durand for bringing this book to HarperCollins. Peter Hubbard is a kind, empathetic, and excellent editor (not mutually exclusive, contrary to popular belief): thank you so much for all your work and support. And thanks and love to Peter McDermott, as always!

The way the brain functions depends on its level of arousal or emotional state (it is "state dependent"). If you are asleep, for example, some brain regions are activated while others are turned off. The same is true no matter what "brain state" or level of "arousal" you are experiencing. In particular, the internal states generated by alarm and threat profoundly affect our capacities for thought and our control over our own behavior.

Table 1 illustrates the shift in controlling area of the brain as an individual moves along the "arousal" continuum, and how that shift can influence style of cognition and caring—or what we call the "sphere of concern" that you are able to consider.

When calm, people can use the most "human" part of their brain, the neocortex, to think abstractly, to consider information already stored in the brain (memories), and to be creative. In this state, you can reflect on the past and plan for the future. It is much easier to "put yourself in someone else's shoes" when you are calm and reflective; your empathic capabilities are at their greatest in this safe, secure mode. Alternatively, however, if someone is in a state of complete terror, his primary focus will be avoiding harm to his own body, his reactions will be driven by lower areas in the brain, and there will be no rational thinking—merely reflexive responses directed toward self-preservation. As people move rightward along the arousal continuum, they become less capable of complex thought and of empathy.

Not surprisingly, when we are worried about our own needs, we have less empathy for the needs of unknown children and families thousands of miles away. We may still feel connected to and have empathy for our neighbors—but the more distressed we become, the more our sphere of concern collapses. Understanding the arousal continuum can help us understand why people are kinder and more considerate when calm—and why severe stress can make us "dumber" and less capable of making the best choices. (Adapted from B. D. Perry, R. A. Pollard, T. L. Blakely, W. L. Baker, and D. Vigilante, "Childhood Trauma, the Neurobiology of Adaptation and 'Use-dependent' Development of the Brain: How 'States' Become 'Traits,'" *Infant Mental Health Journal* 16, no. 4 [1995]: 271–291.)

Groups themselves—families, organizations, communities, and societies—can be seen as biological organisms. Individuals are interconnected by our mirror neurons and the rest of our relational neurobiology; we often share brain states. Consequently, in the same way that one brain's functioning is "state dependent," a group's state is similar. Within any group, happiness, loneliness, and fear are contagious.

Table 2 illustrates how the social and environmental pressures on a group (it could be a family, an organization, a community, or even a whole society) can influence the group's thinking, problem-solving strategies, policies and behaviors toward outsiders, or those low in status within the group.

As with individuals, the thinking styles of a group moves down a gradient under threat. When there are no external threats and resources are plentiful and predictable (column one), most group members will be calm, deliberate, abstract, and rational. This group will have the luxury of thinking in abstract ways to solve any of its current problems. The group will plan for the future and its least powerful members (for example, children if the group is a community or society and employees if the group is a business) can be treated with the most flexible, nurturing, and enriching approaches. When resources become limited and there are economic, environmental, or social threats (column two), groups become less capable of complex, abstract problem solving. Strategies tend to focus on the immediate future (for example, the next funding cycle, the next election cycle, in business, the next quarter—hence "Quarteritis") and all aspects of functioning in the group can regress. The least powerful may be ignored or subject to invasive tactics to shift resources to the leadership or elite.

Table 1: Arousal Continuum/State-Dependent Cognition and Caring

Sphere of Concern	WORLD	COMMUNITY	FAMILY	SELF	BODY INTEGRITY
Sense of Time	Future Past	Days Hours	Hour Minutes	Minutes Seconds	Loss of Sense of Time
PRIMARY/Secondary Brain Area	NEOCORTEX Cortex	CORTEX Subcortex	SUBCORTEX Limbic	LIMBIC Midbrain	MIDBRAIN Brainstem
Cognition	Abstract Creative	Rational Concrete	Emotional Irrational	Reactive	Reflexive
Mental State	CALM	ALERT	ALARM	FEAR	TERROR

In a group, organization, or society under direct threat (column three), the focus of all problem solving becomes the immediate present. Solutions tend to be reactive and regressive. The least powerful are ignored and, if they get in the way, they are even more harshly treated. The less control the group feels it has over the external situation, the more controlling, reactive, and oppressive the internally focused actions of this group will become.

In each of these situations, the prevailing child-rearing styles (or practices directed toward employees or citizens) will create children (or employees or citizens) that will reinforce that group's or society's structure: in a safe and abstract-thinking group the children will be more likely to receive and benefit from enrichment and education, thereby optimizing their potential for creativity, abstraction, and productivity. In contrast, children growing up in groups or societies under threat will be more likely to be raised with harsh or distant caregiving. The result will be impulsive, concrete, and reactive adults, perfectly positioned to fit in

Table 2: State-Dependent Functioning in Groups

Social-Environmental Pressures	Resource-Surplus Predictable Stable/Safe	Resource-Limited Unpredictable Novel	Resource-Poor Inconsistent Threatening
Prevailing Cognitive Style	Abstract Creative	Concrete Superstitious	Reactive Regressive
Prevailing Affective "Tone"	CALM	ANXIETY	TERROR
Systemic Solutions	INNOVATIVE	SIMPLISTIC	REACTIONARY
Focus of Solution	FUTURE	Immediate FUTURE "Quarteritis"	PRESENT
Rules, Regulations, and Laws	Abstract Conceptual	Superstitious Intrusive	Restrictive Punitive
Child-Rearing Styles/Employee Practices	Nurturing Flexible Enriching	Ambivalent Obsessive Controlling	Apathetic Oppressive Harsh

and contribute to a reactive, oppressive, and aggressive group or society. (Adapted from B. D. Perry, "The Neurodevelopmental Impact of Violence in Childhood," *Textbook of Child and Adolescent Forensic Psychiatry,* eds. D. Schetky and E. Benedek [Washington, D.C.: American Psychiatric Press, Inc., 2001], pp. 221–238.)

NOTES

Introduction

1. M. McPherson, L. Smith-Lovin, and M. E. Brashears, "Social Isolation in America: Change in Core Discussion Networks Over Two Decades," *American Sociological Review* 71 (June 2006): 353–375.
2. R. D. Putnam, "Bowling Alone: America's Declining Social Capital," *Journal of Democracy* (January 1995): 65–78, from General Social Survey, 2008, National Opinion Research Center, University of Chicago, www.norc.org.
3. A. Salcedo, T. Schoellman, and M. Tertilt, "Families as Roommates: Changes in U.S. Household Size from 1850 to 2000," *Stanford Institute for Economic Policy Research* (October 2009).

Chapter One: Heaven Is Other People

1. M. Gordon, *Roots of Empathy: Changing the World Child by Child* (Toronto: Thomas Allen Publishers, 2005), pp. 239–252. See also: http://www.rootsofempathy.org/Research.html.
2. A. Paukner, S. J. Suomi, E. Visalberghi, and P. F. Ferrari, "Capuchin Monkeys Display Affiliation Toward Humans Who Imitate Them," *Science* 325, no. 5942 (August 14, 2009): 880–883.
3. G. Rizzolatti and C. Sinigaglia, *Mirrors in the Brain: How Our Minds Share Actions and Emotions* (New York: Oxford University Press, 2008).

4. M. Iacoboni, *Mirroring People: The New Science of How We Connect with Others* (New York: Farrar, Straus Giroux, 2008), p. 110.
5. T. Singer, B. Seymour, J. O'Doherty, H. Kaube, R. J. Dolan, and C. D. Frith, "Empathy for Pain Involves the Affective but Not Sensory Components of Pain," *Science* 303, no. 5661 (February 20, 2004): 1157–1162.
6. A. Smith, *The Theory of Moral Sentiments* (Boston: Wells and Lilly, 1817; repr. 2007), pp. 2–3.

Chapter Two: In Your Face

1. L. Grealy, *Autobiography of a Face* (New York: Harper Perennial, 2003), pp. 6–7.
2. C. Haney, "The Psychological Impact of Incarceration: Implications for Post-Prison Adjustment," *From Prison to Home: The Effect of Incarceration and Reentry on Children, Families, and Communities.* Published online by the U.S. Department of Health and Human Services, December 2001, http://aspe.hhs.gov/HSP/prison2home02/Haney.htm (accessed November 2009).
3. A. Gawande, "Hellhole," *New Yorker,* March 30, 2009, pp. 36–45.
4. M. L. Hoffman, *Empathy and Moral Development: Implications for Caring and Justice* (New York: Cambridge University Press, 2002), p. 199.
5. Ibid.

Chapter Three: Missing People

1. D. Johnson, K. Dole, "International Adoptions: Implications for Early Intervention," *Infants and Young Children* 11, no. 4 (1999): 34–45.
2. U.S. Department of Health and Human Services, Administration for Children & Families, *Child Maltreatment 2007* (Washington, D.C.: U.S. Government Printing Office, 2009), http://www.acf.hhs.gov/programs/cb/pubs/cm07/chapter3.htm#types.
3. D. Iwaniec, *Children Who Fail to Thrive: A Practice Guide* (New York: Wiley, 2004), p. 18.
4. C. A. Nelson, C. H. Zeanah, N. A. Fox, P. J. Marshall, A. T. Smyke, and D. Guthrie, "Cognitive Recovery in Socially Deprived Young

Children: The Bucharest Early Intervention Project," *Science* 318 (December 21, 2007): 1937–1940.

5. Ibid.

6. H. F. Harlow and S. J. Suomi, "Social Recovery by Isolation-Reared Monkeys," *Proceedings of the National Academy of Sciences* 68, no. 7 (July 1, 1971): 1534–1538, http://www.pnas.org/content/68/7/1534. full.pdf.

7. Vole Conference 2009 video; see http://db.cbn.gsu.edu/qtmedia/ Lowell_Getz.mov.

8. Z. R. Donaldson and L. J. Young, "Oxytocin, Vasopressin, and the Neurogenetics of Sociality," *Science* 322 (2008): 900–904, http:// www.sciencemag.org/cgi/content/full/322/5903/900.

9. See http://www.life.illinois.edu/getz/.

10. C. S. Carter, presentation at Vole Conference 2009.

11. P. H. Klopfer, "Mother Love: What Turns It On?" *American Scientist* 59, no. 4 (July 1971): 404–407.

12. T. R. Insel and L. E. Shapiro, "Oxytocin Receptor Distribution Reflects Social Organization in Monogamous and Polygamous Voles," *Proceedings of the National Academy of Sciences* 89, no. 13 (July 1, 1992): 5981–5985, http://www.pnas.org/content/89/13/5981.abstract.

13. A. G. Ophir, S. M. Phelps, A. B. Sorin, and J. O. Wolff, "Social but Not Genetic Monogamy Is Associated with Greater Breeding Success in Prairie Voles," *Animal Behaviour* 75, no. 3 (2008): 1143–1154, doi:10.1016/j.anbehav.2007.09.022.

14. G. Sinha, http://www.gunjansinha.com/popsci_vole.htm.

15. A. B. Wismer Fries, T. E. Ziegler, J. R. Kurian, S. Jacoris, and S. D. Pollak, "Early Experience in Humans Is Associated with Changes in Neuropeptides Critical for Regulating Social Behavior," *Proceedings of the National Academy of Sciences* 102, no. 47 (November 22, 2005): 17237–17240.

Chapter Four: Intense World

1. S. Baron-Cohen, P. Bolton, S. Wheelwright, V. Scahill, L. Short, G. Mead, and A. Smith, "Autism Occurs More Often in Families of Physicists, Engineers, and Mathematicians," *Autism* 2 (1998):

296–301, http://www.autismresearchcentre.com/docs/papers/1998
_B.C.etal_Maths.pdf.

2. "Scientific Brain Linked to Autism," BBC News, January 30, 2006,
http://news.bbc.co.uk/2/hi/health/4661402.stm.

3. S. Baron-Cohen, A. M. Leslie, and U. Frith, "Does the Autistic Child
Have a 'Theory of Mind'?" *Cognition* 21, no. 1 (October 1985): 37–46,
http://ruccs.rutgers.edu/~aleslie/Baron-Cohen%20Leslie%20&
%20Frith%201985.pdf.

4. A. M. Leslie and U. Frith, "Autistic Children's Understanding of
Seeing, Knowing and Believing," *British Journal of Developmental
Psychology* 6 (1988): 315–324.

5. J. M. Flury, W. Ickes, and W. Schweinle, "The Borderline Empathy
Effect: Do High BPD Individuals Have Greater Empathic Ability? Or
Are They Just More Difficult to 'Read'?" *Journal of Research in
Personality* 42, no. 2 (April 2008): 312–332.

6. J. T. Cacioppo and W. Patrick, *Loneliness: Human Nature and the
Need for Social Connection* (New York: Norton, 2008), pp. 216–217.

7. H. Markram, T. Rinaldi, and K. Markram, "The Intense World
Syndrome—An Alternative Hypothesis for Autism," *Frontiers in
Neuroscience* 1, no. 1 (2007): 77–96, http://frontiersin.org/
neuroscience/paper/10.3389/neuro.01/1.1.006.2007/html/.

8. M. F. Casanova, A. E. Switala, J. Trippe, and M. Fitzgerald,
"Comparative Minicolumnar Morphometry of Three Distinguished
Scientists," *Autism* 11, no. 6 (November 2007): 557–569.

9. H. Cody Hazlett, M. Poe, G. Gerig, R. Gimpel Smith, J. Provenzale,
A. Ross, J. Gilmore, and J. Piven, "Magnetic Resonance Imaging and
Head Circumference Study of Brain Size in Autism Birth Through
Age 2 Years," *Archives of General Psychiatry* 62 (2005): 1366–1376.

10. G. Dawson, S. Rogers, J. Munson, M. Smith, et al., "Randomized,
Controlled Trial of an Intervention for Toddlers with Autism: The
Early Start Denver Model," *Pediatrics* (November 30, 2009).

11. I. James, "Singular Scientists," *Journal of the Royal Society of Medicine*
96, no. 1 (January 2003): 36–39, http://www.ncbi.nlm.nih.gov/pmc/
articles/P.M.C539373/?tool=pubmed.

12. W. J. Doherty, *Overscheduled Kids, Underconnected Families: The
Research Evidence,* April 2005, Family Social Science Department,

University of Minnesota, http://www.extension.umn.edu/
parenteducation/research.pdf.
13. V. J. Rideout, E. A. Vandewater, and E. A. Wartella, *Zero to Six: Electronic Media in the Lives of Infants, Toddlers and Preschoolers*, Kaiser Family Foundation, Fall 2003, http://www.kff.org/entmedia/upload/Zero-to-Six-Electronic-Media-in-the-Lives-of-Infants-Toddlers-and-Preschoolers-PDF.pdf.
14. Ibid.

Chapter Five: Lies and Consequences

1. D. G. Rand, A. Dreber, T. Ellingsen, D. Fudenberg, and M. A. Nowak, "Positive Interactions Promote Public Cooperation," *Science* 325, no. 5945 (September 4, 2009): 1272–1275.
2. "Inside the World of Irish Travelers," *Dateline NBC,* October 11, 2002.
3. E. J. Susman, "Psychobiology of Persistent Antisocial Behavior: Stress, Early Vulnerabilities and the Attenuation Hypothesis," *Neuroscience and Biobehavioral Reviews* 30 (2006): 376–389.
4. P. A. Brennan, A. Raine, R. Schulsinger, L. Kirkegaard-Sorensen, J. Knop, B. Hutchings, R. Rosenberg, and S. A. Mednick, "Psychophysiological Protective Factors for Male Subjects at High Risk for Criminal Behavior," *American Journal of Psychiatry* 154 (1997): 853–855.
5. Susman, "Psychobiology of Persistent Antisocial Behavior."
6. T. K. Shackelford, "An Evolutionary Psychological Perspective on Cultures of Honor," *Evolutionary Psychology* 3 (2005): 381–391.
7. R. Hellie, "Interpreting Violence in Late Muscovy from the Perspectives of Modern Neuroscience," Twenty-eighth National Conference of the American Association for the Advancement of Science, Session 7–24, Boston, November 15, 1996.

Chapter Six: No Mercy

1. C. Caldji, J. Diorio, and M. J. Meaney, "Variations in Maternal Care Regulate the Development of Stress Reactivity," *Biological Psychiatry* 48 (2000): 1164–1174.

348   NOTES

2. C. Caldji, B. Tannenbaum, S. Sharma, D. Francis, P. M. Plotsky, and M. J. Meaney, "Maternal Care During Infancy Regulates the Development of Neural Systems Mediating the Expression of Fearfulness in the Rat," *Proceedings of the National Academy of Sciences* 95 (1998): 5335–5340.
3. D. L. Champagne, R. C. Bagot, F. van Hasselt, G. Ramakers, M. J. Meaney, E. R. de Kloet, M. Joëls, and H. Krugers, "Maternal Care and Hippocampal Plasticity: Evidence for Experience-Dependent Structural Plasticity, Altered Synaptic Functioning, and Differential Responsiveness to Glucocorticoids and Stress," *Journal of Neuroscience* 28, no. 23 (June 4, 2008): 6037–6045.
4. N. D. Volkow, G. J. Wang, et al. "Cocaine Cues and Dopamine in Dorsal Striatum: Mechanism of Craving in Cocaine Addiction," *Journal of Neuroscience* 26, no. 24 (June 14, 2006): 6583–6588.
5. F. A. Champagne, P. Chretien, C. W. Stevenson, T. Y. Zhang, A. Gratton, and M. J. Meaney, "Variations in Nucleus Accumbens Dopamine Associated with Individual Differences in Maternal Behavior in the Rat," *Journal of Neuroscience* 24, no. 17 (April 28, 2004): 4113–4123.
6. M. J. Meaney, W. Brake, and A. Gratton, "Environmental Regulation of the Development of Mesolimbic Dopamine Systems: A Neurobiological Mechanism for Vulnerability to Drug Abuse?" *Psychoneuroendocrinology* 27, nos. 1–2 (Jan–Feb 2002): 127–138.
7. D. D. Francis and M. J. Kuhar, "Frequency of Maternal Licking and Grooming Correlates Negatively with Vulnerability to Cocaine and Alcohol Use in Rats," *Pharmacology Biochemistry and Behavior* 90, no. 3 (September 2008): 497–500.
8. V. Engert, R. Joober, M. J. Meaney, D. H. Hellhammer, and J. C. Pruessner, "Behavioral Response to Methylphenidate Challenge: Influence of Early Life Parental Care," *Developmental Psychobiology* 51, no. 5 (July 2009): 408–416.
9. B. Lefkowitz, *Our Guys: The Glen Ridge Rape and the Secret Life of the Perfect Suburb* (New York: Vintage Paperback, 1997), p. 7.
10. Ibid., pp. 155–160, 201–205.
11. Ibid., p. 280.
12. Ibid., p. 278.

Chapter Seven: Resilience

1. M. Iacoboni, *Mirroring People: The New Science of How We Connect with Others* (New York: Farrar, Straus Giroux, 2008), p. 114.
2. D. M. Fergusson and M. T. Lynskey, "Adolescent Resiliency to Family Adversity," *Journal of Child Psychology and Psychiatry, and Allied Disciplines* 37, no. 3 (March 1996): 281–292. See also A. S. Masten, K. B. Burt, G. I. Roisman, J. Obradovic, J. D. Long, and A. Tellegen, "Resources and Resilience in the Transition to Adulthood: Continuity and Change," *Development and Psychopathology* 16, no. 4 (2004): 1071–1094.
3. Intelligence is found repeatedly as a characteristic associated with resilience. See, for example, E. E. Werner, "Risk, Resilience, and Recovery: Perspectives from the Kauai Longitudinal Study," *Development and Psychopathology* 5 (1993): 503–515; Fergusson and Lynskey, "Adolescent Resiliency to Family Adversity," and Masten et al., "Resources and Resilience in the Transition to Adulthood."
4. Werner, "Risk, Resilience, and Recovery."
5. N. Cameron, A. Del Corpo, J. Diorio, K. McAllister, S. Sharma, and M. J. Meaney, "Maternal Programming of Sexual Behavior and Hypothalamic-Pituitary-Gonadal Function in the Female Rat," *PLoS One* 3, no. 5 (May 21, 2008): e2210. (Note: although this is a rat study, it reviews the human research in the same area.)
6. Ibid.
7. M. Wilson and M. Daly, "Life Expectancy, Economic Inequality, Homicide, and Reproductive Timing in Chicago Neighbourhoods," *British Medical Journal* 314 (April 26, 1997): 1271–1274.
8. J. Briere and D. M. Elliott, "Prevalence and Psychological Sequelae of Self-Reported Childhood Physical and Sexual Abuse in a General Population Sample of Men and Women," *Child Abuse & Neglect* 27, no. 10 (October 2003): 1205–1222.
9. V. J. Felitti, R. F. Anda, D. Nordenberg, D. F. Williamson, A. M. Spitz, V. Edwards, M. P. Koss, and J. S. Marks, "Relationship of Childhood Abuse and Household Dysfunction to Many of the Leading Causes of Death in Adults: The Adverse Childhood Experiences (ACE) Study," *American Journal of Preventive Medicine* 14, no. 4 (May 1998): 245–258.

10. D. W. Brown, R. F. Anda, H. Tiemeier, V. J. Felitti, V. J. Edwards, J. B. Croft, and W. H. Giles, "Adverse Childhood Experiences and the Risk of Premature Mortality," *American Journal of Preventive Medicine* 37, no. 5 (November 2009): 389–396.
11. V. J. Felitti et al., op. cit.
12. Ibid.

Chapter Eight: The Chameleon

1. S. A. Powell, C. T. Nguyen, J. Gaziano, V. Lewis, R. F. Lockey, and T. A. Padhya,"Mass Psychogenic Illness Presenting as Acute Stridor in an Adolescent Female Cohort,"*Annals of Otology, Rhinology & Laryngology* 116, no. 7 (July 2007): 525–531.
2. P. H. Hawley, N. Card, and T. D. Little, "The Allure of a Mean Friend: Relationship Quality and Processes of Aggressive Adolescents with Prosocial Skills," *International Journal of Behavioral Development* 31, no. 2 (2007): 170–180.
3. Cited in S. Bowles, "Did Warfare Among Ancestral Hunter-Gatherers Affect the Evolution of Human Social Behaviors," *Science* 324 (June 5, 2009): 1293–1298.
4. Ibid.
5. CIA World Factbook, https://www.cia.gov/library/publications/the-world-factbook/rankorder/2091rank.html?countryName=Angola&countryCode=AO&regionCode=af#AO.
6. S. Hrdy, *Mothers and Others: The Evolutionary Origins of Mutual Understanding* (Cambridge, MA: Belknap Press, 2009), p. 108.
7. Ibid., p. 75.
8. Ibid., p. 108.
9. Ibid., p. 103.
10. Ibid., p. 28.
11. S. E. Asch, "Effects of Group Pressure upon the Modification and Distortion of Judgment," in *Groups, Leadership and Men,* ed. H. Guetzkow (Pittsburgh, PA: Carnegie Press, 1951).
12. S. E. Asch, "Opinions and Social Pressure," *Scientific American* 193, no. 5 (1955): 31–35.

13. T. Blass, "The Man Who Shocked the World," *Psychology Today* 35, no. 2 (March/April 2002): 68–74.
14. T. Blass, "Understanding Behavior in the Milgram Obedience Experiment: The Role of Personality, Situations, and Their Interactions," *Journal of Personality and Social Psychology* 60 (1991): 398–413.
15. S. Burnett, G. Bird, J. Moll, C. Frith, and S. J. Blakemore, "Development During Adolescence of the Neural Processing of Social Emotion," *Journal of Cognitive Neuroscience* 21, no. 9 (September 2009): 1736–1750.
16. X. Xu, X. Zuo, X. Wang, and S. Han, "Do You Feel My Pain? Racial Group Membership Modulates Empathic Neural Responses," *Journal of Neuroscience* 29 (2009): 8525–8529.
17. R. Kurzban, J. Tooby, and L. Cosmides, "Can Race Be Erased? Coalitional Computation and Social Categorization," *Proceedings of the National Academy of Sciences* 98, no. 26 (December 18, 2001): 15,387–15,392.
18. M. Bond, "Critical Mass," *New Scientist,* July 18, 2009, citing M. Levine, A. Prosser, D. Evans, and S. Reicher, "Identity and Emergency Intervention: How Social Group Membership and Inclusiveness of Group Boundaries Shape Helping Behavior," *Personality and Social Psychology Bulletin* 31 (2009): 443–453.
19. P. Bronson, A. Merryman, "Teens Who Feel More Peer Pressure Turn Out Better, Not Worse," *Newsweek,* NurtureShock blog, September 23, 2009.
20. M. T. Moreira, L. A. Smith, and D. Foxcroft, "Social Norms Interventions to Reduce Alcohol Misuse in University or College Students," *Cochrane Database Systematic Reviews* 3 (July 2009).
21. S. W. Henggeler, M. D. Rowland, C. Halliday-Boykins, et al., "One-Year Follow-Up of Multisystemic Therapy as an Alternative to the Hospitalization of Youths in Psychiatric Crisis," *Journal of the American Academy of Child and Adolescent Psychiatry* 42, no. 5 (May 2003): 543–551.
22. N. A. Christakis and J. H. Fowler, "The Spread of Obesity in a Large Social Network over 32 Years," *New England Journal of Medicine* 357, no. 4 (July 2007): 370–379.

23. Ibid., "The Collective Dynamics of Smoking in a Large Social Network," *New England Journal of Medicine* 358, no. 21 (May 2008): 2249–2258; "Dynamic Spread of Happiness in a Large Social Network: Longitudinal Analysis over 20 Years in the Framingham Heart Study," *BMJ* 337, no. a2338 (December 4, 2008); J. T. Cacioppo, J. H. Fowler, and N. A. Christakis, "Alone in the Crowd: The Structure and Spread of Loneliness in a Large Social Network," *Journal of Personality and Social Psychology* (pre-published December 2009).

24. P. M. Oliner and S. Oliner, *Toward a Caring Society: Ideas into Action* (Westport, CT: Praeger, 1995), p. 6.

25. U. Gatti, R. E. Tremblay, and F. Vitaro, "Iatrogenic Effect of Juvenile Justice," *Journal of Child Psychology and Psychiatry* 50, no. 8 (August 2009): 991–998.

Chapter Nine: Us Versus Them

1. S. Levitt, "Steven Levitt Analyzes Crack Economics," TED Talks, February 2004, http://www.ted.com/talks/steven_levitt_analyzes_crack_economics.html (accessed November 2009).

2. Risley and Hart, cited in P. Tough, "What It Takes to Make a Student," *New York Times,* November 26, 2006, http://www.nytimes.com/2006/11/26/magazine/26tough.html.

3. R. D. S. Raizada, T. L. Richards, A. Meltzoff, and P. K. Kuhl, "Socioeconomic Status Predicts Hemispheric Specialisation of the Left Inferior Frontal Gyrus in Young Children," *NeuroImage* 40, no. 3 (April 15, 2008): 1392–1401.

4. B. P. Marx, K. Brailey, S. P. Proctor, H. Z. Macdonald, et al., "Association of Time Since Deployment, Combat Intensity, and Posttraumatic Stress Symptoms with Neuropsychological Outcomes Following Iraq War Deployment," *Archives of General Psychiatry* 66, no. 9 (September 2009): 996–1004.

5. T.-N. Coates, *The Beautiful Struggle: A Father, Two Sons and an Unlikely Road to Manhood* (New York: Spiegel and Grau, 2009), p. 177.

6. Ibid., pp. 148–149.

7. S. S. Wiltermuth and C. Heath, "Synchrony and Cooperation," *Psychological Science* 20, no. 1 (January 2009): 1–5.

8. R. Hahn, A. McGowan, A. Liberman, et al., "Effects on Violence of Laws and Policies Facilitating the Transfer of Youth from the Juvenile to the Adult Justice System," *Morbidity and Mortality Weekly Report* 56 (2007): 1–11, http://www.cdc.gov/mmwr/preview/mmwrhtml/rr5609a1.htm.

Chapter Ten: Glued to the Tube

1. Agency for Health Care Research and Quality, *Perinatal Depression Prevalence, Screening Accuracy, and Screening Outcomes Summary Evidence Report/Technology Assessment: Number 119,* 2005, http://www.ahrq.gov/clinic/epcsums/peridepsum.htm (accessed October 2009).
2. E. H. Hagen, "The Functions of Postpartum Depression," *Evolution and Human Behavior* 20, no. 5 (September 1999): 325–359.
3. A. Stein, L. E. Malmberg, K. Sylva, J. Barnes, P. Leach P, and the FCCC team, "The Influence of Maternal Depression, Caregiving, and Socioeconomic Status in the Post-Natal Year on Children's Language Development," *Child Care Health and Development* 34, no. 5 (September 2008): 603–612.
4. J. Ham and E. Tronick, "Infant Resilience to the Stress of the Still-Face: Infant and Maternal Psychophysiology Are Related," *Annals of the New York Academy of Sciences* 1094 (December 2006): 297–302.
5. A. Huston, E. Donnerstein, H. Fairchild, N. Feshbach, P. Katz, J. Murray, E. Rubinstein, B. Wilcox, and D. Zuckerman, *Big World, Small Screen: The Role of Television in American Society* (Lincoln: University of Nebraska Press, 1992).
6. L. R. Huesmann and L. D. Taylor, "The Role of Media Violence in Violent Behavior," *The Annual Review of Public Health* 27 (2006): 393–415.
7. L. R. Huesmann, J. Moise-Titus, C.-L. Podolski, and L. D. Eron, "Longitudinal Relations Between Children's Exposure to TV Violence and Their Aggressive and Violent Behavior in Young Adulthood: 1977–1992," *Developmental Psychology* 39, no. 2 (2003): 201–221, http://www.apa.org/journals/releases/dev392201.pdf.
8. S. Levitt and S. Dubner, *SuperFreakonomics* (New York: William Morrow, 2009), p. 104.

9. J. M. Ostrov, D. A. Gentile, and N. R. Crick, "Media Exposure, Aggression and Prosocial Behavior During Early Childhood: A Longitudinal Study," *Social Development* 15, no. 4 (October 2006): 612–627, http://drdouglas.org/drdpdfs/Ostrov_Gentile_Crick_in _press.pdf.

10. L. R. Huesmann and L. D. Taylor, "The Role of Media Violence in Violent Behavior," *Annual Review of Public Health* 27 (April 2006): 393–415.

11. B. W. Newton, L. Barber, J. Clardy, E. Cleveland, and P. O'Sullivan, "Is There Hardening of the Heart During Medical School?" *Academic Medicine: Journal of the Association of American Medical Colleges* 83, no. 3 (March 2008): 244–249.

12. Huesmann et al., "Longitudinal Relations."

13. G. Orwell, *My Country Right or Left, 1940–43: The Collected Essays, Journalism & Letters of George Orwell,* eds. Sonia Orwell and Ian Angus (New York: Harcourt, Brace, and World, 1968; repr. Jaffrey, NH: Nonpareil Books, 2000), p. 254.

14. J. Glover, *Humanity: A Moral History of the Twentieth Century* (New Haven, CT: Yale University Press, 1999), pp. 159–160.

15. B. J. Bushman and C. A. Anderson, "Comfortably Numb: Desensitizing Effects of Violent Media on Helping Others," *Psychological Science* 20, no. 3 (2008): 273–277, http://sitemaker .umich.edu/brad.bushman/files/ba09.pdf.

16. A. I. Nathanson et al., "Effects of Parental TV Restrictions Among a Group of Teens," *Media Psychology,* November 2002; A. I. Nathanson, M. S. Yang, "The Effects of Mediation Content and Form on Children's Responses to Violent Television," *Human Communication Research* 29, no. 1 (January 2003): 111–134.

17. L. R. Huesmann and L. D. Taylor, op. cit.

18. R. Putnam, *Bowling Alone: The Collapse and Revival of American Community* (New York: Simon & Schuster, 2000).

19. Ibid., p. 242.

20. Ibid., pp. 238–240; S. Sidney, B. Sternfeld, W. L. Haskell, D. R. Jacobs Jr., M. A. Chesney, and S. B. Hulley, "Television Viewing and Cardiovascular Risk Factors in Young Adults: The CARDIA Study," *Annals of Epidemiology* 6, no. 2 (March 1996): 154–159.

21. B. J. Bushman, "Does Venting Anger Feed or Extinguish the Flame? Catharsis, Rumination, Distraction, Anger, and Aggressive Responding," *Personality and Social Psychology Bulletin* 28, no. 6 (2002): 724–731.

22. D. Elkind, *The Power of Play* (Cambridge, MA: Da Capo Books, 2007).

23. S. M. Lee, C. R. Burgeson, J. E. Fulton, and C. G. Spain, "Physical Education and Physical Activity: Results from the School Health Policies and Programs Study 2006," *Journal of School Health* 77, no. 8 (October 2007): 435–463, http://www.ashaweb.org/files/public/JOSH_1007/josh_77_8_lee_p_435.pdf.

Chapter Eleven: On Baboons, British Civil Servants, and the Oscars

1. M. Marmot, *The Status Syndrome* (New York: Henry Holt, 2004), p. 22.

2. M. Bond, "Why Cops Should Trust the Wisdom of the Crowds," *New Scientist* 2717 (July 17, 2009).

3. M. Marmot, *The Status Syndrome*, p. 39.

4. R. G. Wilkinson, *Mind the Gap* (New Haven, CT: Yale University Press, 2001), p. 5.

5. A. Sacker, D. Firth, R. Fitzpatrick, K. Lynch, and M. Bartley, "Comparing Health Inequality in Men and Women: Prospective Study of Mortality 1986–96," *British Medical Journal* 320, no. 7245 (May 13, 2000): 1303–1307; also see Marmot, *The Status Syndrome*.

6. R. M. Sapolsky, *Why Zebras Don't Get Ulcers* (New York: W. H. Freeman, 1998, reprint edition by Barnes and Noble), p. 291.

7. Marmot, *The Status Syndrome*, pp. 44–45.

8. Sapolsky, *Zebras*, pp. 294–295.

9. R. M. Sapolsky and L. J. Share, "A Pacific Culture among Wild Baboons: Its Emergence and Transmission," *PLoS Biology* 2, no. 4 (April 2004): 0534–0541.

10. B. A. Arnow, "Relationships between Childhood Maltreatment, Adult Health and Psychiatric Outcomes, and Medical Utilization," *Journal of Clinical Psychiatry* 65, suppl. 12 (2004): 10–15.

11. V. Krishnan and E. J. Nestler, "The Molecular Neurobiology of Depression," *Nature* 455, no. 7215 (October 16, 2008): 894–902.

12. P. Rohde, "The Relevance of Hierarchies, Territories, Defeat for Depression in Humans: Hypotheses and Clinical Predictions," *Journal of Affective Disorders* 65, no. 3 (August 2001): 221–230.
13. M. J. Raleigh, M. T. McGuire, G. L. Brammer, D. B. Pollack, and A. Yuwiler, "Serotonergic Mechanisms Promote Dominance Acquisition in Adult Male Vervet Monkeys," *Brain Research* 559, no. 2 (September 1991): 181–190.
14. M. Seligman, with K. Reivich, L. Jaycox, and J. Gillham, *The Optimistic Child* (Boston: Houghton Mifflin, 1995), pp. 2–4.
15. J. E. Malberg and R. S. Duman, "Cell Proliferation in Adult Hippocampus Is Decreased by Inescapable Stress: Reversal by Fluoxetine Treatment," *Neuropsychopharmacology* 28, no. 9 (September 2003): 1562–1571.
16. M. A. Visintainer, J. R. Volpicelli, and M. E. Seligman, "Tumor Rejection in Rats after Inescapable or Escapable Shock," *Science* 216, no. 4544 (April 23, 1982): 437–439; P. A. Rittenhouse, C. López-Rubalcava, G. D. Stanwood, and I. Lucki, "Amplified Behavioral and Endocrine Responses to Forced Swim Stress in the Wistar-Kyoto Rat," *Psychoneuroendocrinology* 27, no. 3 (April 2002): 303–318.

Chapter Twelve: Warm as Iceland

1. United Nations Development Program, *United Nations Human Development Index, 2008,* http://hdr.undp.org/en/media/HDI_2008_EN_Complete.pdf (accessed November 2009).
2. Vision of Humanity website, http://www.visionofhumanity.org/gpi/results/iceland/2008.
3. United Nations Office on Drugs and Crime, *The Ninth United Nations Survey on Crime Trends and the Operations of Criminal Justice Systems (2003–2004),* http://www.unodc.org/documents/data-and-analysis/CTS9_by_country_public.pdf (accessed November 2009).
4. United States Bureau of Justice Statistics, *Prison Statistics,* June 30, 2008, http://www.ojp.usdoj.gov/bjs/prisons.htm (accessed November 2009).
5. J. Carlin, "No Wonder Iceland Has the Happiest People on Earth," *Observer,* May 18, 2008.

6. R. Wade, "The Crisis: Iceland as Icarus," *Challenge* 52, no. 3 (May/June 2009): 5–33.

7. G. T. Svendsen and G. L. H. Svendsen, eds., *Handbook of Social Capital: The Troika of Sociology, Political Science and Economics* (Northampton, MA: Edward Elgar, 2009), pp. 7–10.

8. Ibid.

9. A. Antoci, F. Sabatini, and M. Sodini, "The Fragility of Social Capital," AICCON Working Papers 59, Associazione Italiana per la Cultura della Cooperazione e del Non Profit (2009).

10. M. Kosfeld, M. Heinrichs, P. J. Zak, U. Fischbacher, and E. Fehr, "Oxytocin Increases Trust in Humans," *Nature* 435 (2005): 673–676.

11. P. Zak, A. Stanton, and S. Ahmadi, "Oxytocin Increases Generosity in Humans," *PloS One* 2, no. 11 (2007): e1128, http://www.plosone.org/article/info:doi/10.1371/journal.pone.0001128.

12. T. Baumgartner, M. Heinrichs, A. Vonlanthen, U. Fischbacher, and E. Fehr, "Oxytocin Shapes the Neural Circuitry of Trust and Trust Adaptation in Humans," *Neuron* 58, no. 4 (May 22, 2008): 639–650.

13. P. J. Zak, R. Kurzband, and W. T. Matznere, "Oxytocin Is Associated with Human Trustworthiness," *Hormones and Behavior* 48, no. 5 (December 2005): 522–527.

14. P. J. Zak, A. A. Stanton, and S. Ahmadi, "Oxytocin Increases Generosity in Humans," *PLoS One* 2, no. 11 (November 2007): e1128.

15. T. Nørretranders, *The Generous Man: How Helping Others Is the Sexiest Thing You Can Do* (New York: Avalon, 2005), p. 6.

16. A. Serino, G. Giovagnoli, and E. Làdavas, "I Feel What You Feel If You Are Similar to Me," *PLoS ONE* 4, no. 3 (March 2009): e4930.

17. B. Moyers, "Wendell Potter on Profits Before Patients," *Bill Moyers Journal*, PBS, July 31, 2009, http://www.pbs.org/moyers/journal/07102009/profile.html (accessed November 2009).

18. Ibid.

19. Ibid.

20. D. Wessel, "Escalator Ride: As Rich-Poor Gap Widens in the U.S., Class Mobility Stalls," *Wall Street Journal*, May 13, 2005.

21. All Headline News, "OECD Report Finds Rising Income Inequality with U.S. Among Worst," October 21, 2008, http://www.allheadlinenews.com/articles/7012734983.

22. E. Saez, "Striking It Richer: The Evolution of Top Incomes in the United States," MIT Department of Economics, http://emlab. berkeley.edu/users/saez/saez-UStopincomes-2007.pdf.

23. C. R. Morris, "Freakoutonomics," *New York Times,* June 2, 2006, http://www.nytimes.com/2006/06/02/opinion/02morris.html.

24. The Century Foundation, *Divided We Fail: Coming Together through Public School Choice* (New York: The Century Foundation Press, 2002), p.15, http://www.tcf.org/Publications/Education/ dividedwefail.pdf.

25. "What Do Americans Really Want? TV and Sex," September 15, 2009, http://today.msnbc.msn.com/id/32862930/ns/today-today_ books.

26. United States: FBI Uniform Crime Reports, 2004, http://www.fbi .gov/ucr/cius_04/offenses_reported/violent_crime/murder.html (accessed November 2009).
    Iceland: United Nations Office on Drugs and Crime, op. cit. Note: 2004 statistics used here because these are the latest available for Iceland.

27. United States: P. Harrison and A. Beck, "Prison and Jail Inmates at Midyear 2004," *Bureau of Justice Statistics Bulletin,* http://www.ojp. usdoj.gov/bjs/pub/pdf/pjim04.pdf (accessed November 2009).
    Iceland: Ibid.

28. R. G. Wilkinson, *Mind the Gap* (New Haven, CT: Yale University Press, 2001), p. 15.

29. Transparency International website, http://www.transparency.org/ news_room/in_focus/2008/cpi2008/cpi_2008_table.

30. Index Mundi website, http://www.indexmundi.com/g/r.aspx?c=ic& v=30.

31. Iceland: CIA World Factbook, https://www.cia.gov/library/publications/ the-world-factbook/geos/ic.html (accessed November 2009).
    United States: CIA World Factbook, https://www.cia.gov/library/ publications/the-world-factbook/geos/us.html (accessed November 2009).

32. Wilkinson, *Mind the Gap,* p. 5; see also Whitehall I, cited in Michael Marmot, *The Status Syndrome* (New York: Henry Holt, 2005).

33. Statistics Iceland, http://www.statice.is/?PageID=444&NewsID=3776 (accessed November 2009).

34. G. B. Eydal and S. Ólafsson, "Demographic Trends in Iceland," First report for the project *Welfare Policy and Employment in the Context of Family Change* (May 2003).
35. J. Belsky, D. Lowe Vandell, M. Burchinal, K. A. Clarke-Stewart, et al., "Are There Long-Term Effects of Early Child Care?" *Child Development* 78, no. 2 (March/April 2007): 681–701.
36. C. Koverola, M. A. Papas, S. Pitts, C. Murtaugh, M. M. Black, and H. Dubowitz, "Longitudinal Investigation of the Relationship Among Maternal Victimization, Depressive Symptoms, Social Support, and Children's Behavior and Development," *Journal of Interpersonal Violence* 20, no. 12 (December 2005): 1523–1546; B. L. Green, C. Furrer, and C. McAllister, "How Do Relationships Support Parenting? Effects of Attachment Style and Social Support on Parenting Behavior in an At-Risk Population," *American Journal of Community Psychology* 40, nos. 1–2 (September 2007): 96–108.
37. R. Wade, "The Crisis," op. cit.
38. Iceland: "December Unemployment Figures," *IceNews,* January 16, 2009, http://www.icenews.is/index.php/2009/01/16/iceland -december-unemployment-figures/; October 2009 numbers: http://www.icelandreview.com/icelandreview/search/news/Default.asp ?ew_0_a_id=351863.
United States: Regional and State Employment and Unemployment Summary, Economic News Release, Bureau of Labor Statistics, http://www.bls.gov/news.release/laus.nr0.htm.

Chapter Thirteen: All Together Now

1. P. Arnstein, M. Vidal, C. Wells-Federman, B. Morgan, and M. Caudill, "From Chronic Pain Patient to Peer: Benefits and Risks of Volunteering," *Pain Management Nursing* 3, no. 3 (September 2002): 94–103.
2. D. Oman, C. E. Thoresen, and K. McMahon, "Volunteerism and Mortality Among the Community-Dwelling Elderly," *Journal of Health Psychology* 4, no. 3 (1999): 301–316.
3. L. Scherwitz, R. McKelvain, C. Laman, J. Patterson, L. Dutton, S. Yusim, J. Lester, I. Kraft, D. Rochelle, and R. Leachman, "Type A

Behavior, Self-Involvement, and Coronary Atherosclerosis,"
Psychosomatic Medicine 45, no. 1 (1983): 47–57.

4. R. Putnam, *Bowling Alone.*

5. Ibid; and for 2008, General Social Survey.

6. R. Roth, *American Homicide* (Cambridge, MA: Belknap Press of
Harvard University Press, 2009).

7. U.S. Census, *Families and Living Arrangements,* 2008, http://www.
census.gov/population/www/socdemo/hh-fam.html (accessed
November 2009).

8. S. Coontz, *The Way We Never Were: American Families and the
Nostalgia Trap* (New York: Basic Books, 1992, repr. 2000), p. 15.

9. D. Cohn and R. Morin, "American Mobility: Who Moves? Who Stays
Put? Where's Home?" Pew Research Center, December 2008, http://
pewsocialtrends.org/assets/pdf/Movers-and-Stayers.pdf (accessed
November 2009).

10. Ibid.

11. A. Salcedo, T. Schoellman, and M. Tertilt, *Families as Roommates:
Changes in U.S. Household Size from 1850 to 2000,* Stanford Institute
for Economic Policy Research (October 2009).

12. M. McPherson, L. Smith-Lovin, and M. E. Brashears, "Social
Isolation in America: Change in Core Discussion Networks Over
Two Decades," *American Sociological Review* 71 (June 2006):
353–375.

13. Ibid.

14. B. Stevenson and J. Wolfers, "Marriage and Divorce: Changes and
their Driving Forces," *Journal of Economic Perspectives* 21, no. 2
(Spring 2007): 27–52.

15. U.S. Census Bureau, "Number, Timing, and Duration of Marriages
and Divorces: 2001," February 2005, http://www.census.gov/prod/
2005pubs/p70-97.pdf.

16. Children and Nature Network, "What the Research Shows,"
http://www.childrenandnature.org/reports/8_2007/resourcestools/
WhattheResearchShows.pdf (accessed November 2009).

17. W. J. Doherty, "Overscheduled Kids, Underconnected Families:
The Research Evidence," Family Social Science Department,
University of Minnesota, April 2005, http://www.shouldertoshoulder

minnesota.org/files/Overscheduled%20Kids,%20Underconnected%20
Families%20(Research%20Evidence).doc (accessed November 2009).

18. S. M. Lee, C. R.Burgeson, et al., "Physical Education and Physical
Activity: Results from the School Health Policies and Programs
Study 2006," *Journal of School Health* 77, no. 8 (October 2007):
435–463, http://www.ashaweb.org/files/public/JOSH_1007/
josh_77_8_lee_p_435.pdf (accessed November 2009).

19. BJS statistics 2006, table 1.29, cited in E. Anderson, ed., *Against the
Wall: Poor, Young, Black, and Male* (Philadelphia: University of
Pennsylvania Press, 2008), p. 72.

20. L. E. Glaze and L. M. Maruschak, "Parents in Prison and Their
Minor Children," Bureau of Justice Statistics Special Report, updated
January 8, 2009, http://www.ojp.usdoj.gov/bjs/pub/pdf/pptmc.pdf
(accessed November 2009).

21. Senator Jim Webb's floor speech to introduce the National Criminal
Justice Commission Act of 2009, November 3, 2009, http://webb.
senate.gov/newsroom/pressreleases/2009-11-03-02.cfm (accessed
November 2009).

22. A. Gawande, "Hellhole," p. 42.

23. V. J. Rideout, E. A. Vandewater, and E. A. Wartella, "Zero to Six:
Electronic Media in the Lives of Infants, Toddlers and Preschoolers,"
Henry J. Kaiser Family Foundation report, Fall 2003, http://www.kff
.org/entmedia/upload/Zero-to-Six-Electronic-Media-in-the-Lives-of-
Infants-Toddlers-and-Preschoolers-PDF.pdf (accessed November 2009).

24. Ibid.

25. S. M. Bianchi, J. P. Robinson, and M. A. Milkie, *Changing Rhythms of
American Family Life* (New York: Russell Sage Foundation, 2006).

26. Ibid.

27. M. Foucault, *Discipline and Punish: The Birth of the Prison* (New
York: Vintage, 1995), p. 1.

28. S. Pinker, "A History of Violence" *Edge,* http://www.edge.org/3rd_
culture/pinker07/pinker07_index.html (accessed November 2009).

29. M. Eisner, "The Long-Term Development of Violence: Empirical
Findings and Theoretical Approaches to Interpretation,"
International Handbook of Violence Research, eds. W. Heitmayer and
J. Hagan (Boston: Kluwer Academic Publishers, 2003), p. 45.

30. FBI, Crime in the United States, table 1, 2008, http://www.fbi.gov/
    ucr/cius2008/data/table_01.html (accessed November 2009).
31. UNICEF, "Child Poverty in Rich Countries, 2005," *Innocenti Report
    Card No. 6.*, UNICEF Innocenti Research Centre, Florence,
    http://www.unicef.gr/reports/rc06/UNICEF%20CHILD%20POV
    ERTY%20IN%20RICH%20COUNTRIES%202005.pdf (accessed
    November 2009).
32. P. Kulhara, R. Shah, and S. Grover, "Is the Course and Outcome of
    Schizophrenia Better in the 'Developing' World?" *Asian Journal of
    Psychiatry* 2, no. 2 (June 2009): 55–62.
33. A. Phillips and B. Taylor, *On Kindness* (New York: Farrar, Straus and
    Giroux, 2009), pp. 8–10.
34. W. Ickes, *Everyday Mind Reading: Understanding What Other People
    Think and Feel* (Amherst, NY: Prometheus Books, 2003), p. 131.
35. M. A. Straus and M. J. Paschall, "Corporal Punishment by Mothers
    and Development of Children's Cognitive Ability: A Longitudinal
    Study of Two Nationally Representative Age Cohorts," *Journal
    of Aggression, Maltreatment & Trauma* 18, no. 5 (July 2009): 459–483.
36. J. E. Lansford, M. M. Criss, K. A.Dodge, D. S. Shaw, G. S. Pettit, and
    J. E. Bates, "Trajectories of Physical Discipline: Early Childhood
    Antecedents and Developmental Outcomes," *Child Development* 80,
    no. 5 (September/October 2009): 1385–1402.
37. S. Bennett and N. Kalish, *The Case Against Homework: How
    Homework Is Hurting Our Children and What We Can Do About It*
    (New York: Crown Publishers, 2006).
38. A. Lareau, *Unequal Childhoods: Class, Race, and Family Life*
    (Berkeley: University of California Press, 2003). See especially
    pp. 241–257.
39. R. Blum, "School Connectedness: Improving Students' Lives,"
    Journal of School Health, September 2004, http://cecp.air.org/
    download/MCMonographFINAL.pdf (accessed November 2009).
40. M. Gladwell, *Outliers: The Story of Success* (New York: Little, Brown
    and Co., 2008), pp. 3–11.
41. S. Wolf and J. G. Bruhn, *The Power of Clan: The Influence of Human
    Relationships on Heart Disease* (New Brunswick, NJ: Transaction
    Publishers, 1998), pp. 10–11.

Epilogue: People and Programs

1. R. Shulman, "Harlem Program Singled Out as Model; Obama Administration to Replicate Plan in Other Cities to Boost Poor Children," *Washington Post*, August 2, 2009.
2. P. Tough, *Whatever It Takes: Geoffrey Canada's Quest to Change Harlem and America* (New York: Mariner Books, 2009), p. 80.
3. Ibid., p. 81.
4. Shulman, op. cit.
5. Ibid.
6. D. Johnson, "Rantoul Journal: For Distant Generations in Illinois, Unrelated but Oh So Close," *New York Times,* September 15, 2008, http://www.nytimes.com/2008/09/16/us/16rantoul.html?_r=1&em.

INDEX